INTERNATIONAL ENERGY AGENCY

The link between ENERGY & HUMAN ACTIVITY

INTERNATIONAL ENERGY AGENCY

9, rue de la Fédération,
75739 Paris Cedex 15, France

The International Energy Agency (IEA) is an autonomous body which was established in November 1974 within the framework of the Organisation for Economic Co-operation and Development (OECD) to implement an international energy programme.

It carries out a comprehensive programme of energy co-operation among twenty four* of the OECD's twenty nine Member countries.
The basic aims of the IEA are:

- To maintain and improve systems for coping with oil supply disruptions;
- To promote rational energy policies in a global context through co-operative relations with non-Member countries, industry and international organisations;
- To operate a permanent information system on the international oil market;
- To improve the world's energy supply and demand structure by developing alternative energy sources and increasing the efficiency of energy use;
- To assist in the integration of environmental and energy policies.

**IEA Member countries: Australia, Austria, Belgium, Canada, Denmark, Finland, France, Germany, Greece, Hungary, Ireland, Italy, Japan, Luxembourg, the Netherlands, New Zealand, Norway, Portugal, Spain, Sweden, Switzerland, Turkey, the United Kingdom, the United States. The European Commission also takes part in the work of the IEA.*

ORGANISATION FOR ECONOMIC CO-OPERATION AND DEVELOPMENT

Pursuant to Article 1 of the Convention signed in Paris on 14th December 1960, and which came into force on 30th September 1961, the Organisation for Economic Co-operation and Development (OECD) shall promote policies designed:

- To achieve the highest sustainable economic growth and employment and a rising standard of living in Member countries, while maintaining financial stability, and thus to contribute to the development of the world economy;
- To contribute to sound economic expansion in Member as well as non-Member countries in the process of economic development; and
- To contribute to the expansion of world trade on a multilateral, non-discriminatory basis in accordance with international obligations.

The original Member countries of the OECD are Austria, Belgium, Canada, Denmark, France, Germany, Greece, Iceland, Ireland, Italy, Luxembourg, the Netherlands, Norway, Portugal, Spain, Sweden, Switzerland, Turkey, the United Kingdom and the United States. The following countries became Members subsequently through accession at the dates indicated hereafter: Japan (28th April 1964), Finland (28th January1969), Australia (7th June 1971), New Zealand (29th May 1973), Mexico (18th May 1994), the Czech Republic (21st December 1995), Hungary (7th May 1996), Poland (22nd November 1996) and the Republic of Korea (12th December 1996). The Commission of the European Communities takes part in the work of the OECD (Article 13 of the OECD Convention).

Foreword

Every day millions of people make decisions which determine how energy is used. They commute to school and work, produce goods and render services, haul freight, heat their homes and offices. Energy serves as a means to these ends. And the ends define the proper study of energy use and of the CO_2 emissions it generates. Energy consumption has its roots in the ways economies and societies work.

This book distills for general readers the essential messages of the International Energy Agency's recently published Indicators of Energy Use and Efficiency, *which examines the development of end-use "energy indicators" and their relevance to policy-making. This volume offers a developing system of relatively straightforward indicators. It demonstrates how their use can effectively contribute to understanding the complex fabric of energy demand in industrialised countries.*

Built-up from disaggregated data, the indicators provide a link between bottom-up and top-down approaches to examine the evolution of energy end-use. The underlying data come from a variety of national and international sources and may not be uniform. What is new is how they are gathered and combined in the IEA secretariat to produce data and indicators which are, so far as possible, compatible among countries.

Indicators also establish quantitative links between energy end-use patterns and CO_2 emissions. While IEA Member countries significantly shrunk the amount of carbon they released per unit of production between 1973 and 1993, that reduction has slowed dramatically. Carbon emissions in total are rising, not falling. Strategies for future reduction of carbon emissions are unlikely to succeed unless policy makers watch the key trends in energy end-use. Indicators such as those summarised here reveal and explain those trends and the forces driving them.

Robert Priddle, Executive Director

Acknowledgements

This book is drawn substantively from the IEA's *Indicators of Energy Use and Efficiency* whose author, Dr Lee Schipper, crafted it from numerous studies from the Member countries of the IEA. The contribution of the many institutions and individuals provides the foundation of the work presented here and is gratefully acknowledged. In particular, thanks go to the Energy Analysis Program of Lawrence Berkeley National Laboratory and the Fraunhofer Institut for timely assistance in updating the statistics.

The main contributors to this volume include Lee Schipper, Jean-Pierre DesRosiers, Debra Justus, Robert Cornell and Scott Sullivan. The graphic and production assistance of Muriel Custodio and Corinne Hayworth added significantly to the material presented and is gratefully acknowledged.

Hans Jørgen Koch

Director, Office of Energy Efficiency, Technology, Research and Development

November 1997

TABLE OF CONTENTS

List of Figures

The Purpose and Scope of Energy Indicators

OVERVIEW

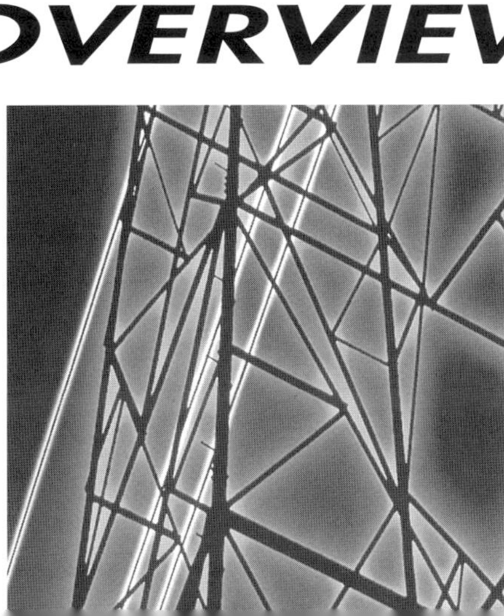

1

Who's in Charge Here?

■ There are hundreds of millions of them: households and car drivers; millions of truckers; hundreds of thousands of building operators; farmers and factory managers. These are the people, the armies of discrete individuals, who make the decisions that govern energy use and CO_2 emissions.

■ Each day, they make billions of energy choices to achieve their manifold goals, few of which have much to do with energy itself. What they are interested in is producing goods, rendering services, heating houses, driving to work and hauling freight, to name just a few. Some energy applications generate output and income. Others provide convenience or comfort, or simply facilitate the business of life. Energy serves as a means to these ends. The ends themselves define the proper study of how and why people burn hydrocarbons and ultimately release carbon dioxide into the atmosphere. Energy consumption has its roots in the ways economies and societies work.

Policies: the Devils Are in the Details

■ Political authorities and other policy makers do not make the crucial decisions about energy use. Yet they play a vital role in guiding the individual choices which affect energy consumption and CO_2 emissions. For free markets do not take fully into account the environmental consequences of those choices and their impact on future generations. How can the policy makers develop programmes which work, which provide effective guidance for the millions of actual decision makers and enforce that guidance in ways that are politically acceptable?

■ Rational programme design depends vitally on information – on data which emerge from the grass roots decision processes and which accurately describe both these processes and their effects. The policy maker needs both good data and the right kind of data.

■ But what is the "right kind" of data? For decades, most energy policy analysis relied on highly aggregated statistics; such analysis concentrated on the correlation over time and among countries between overall energy demand and economic activity as measured by GDP. This aggregate approach proved moderately useful so long as it remained possible to tell the time from the clock face without asking much about the machinery inside – before the oil price shocks of the 1970's shattered the simple message drawn from that correlation. It shattered the adequacy of the correlation for policy-making from a time when the main concern was how much supply would be needed. Analysts now recognise that, while energy demand rises with economic growth almost everywhere, the really significant story lies in how this coupling varies from sector to sector, from country to country and from period to period. Equally important is an understanding of how energy users respond to a host of variable factors with specific micro-impacts: income; energy prices; technology; energy efficiency; structural changes in the mix of goods and services demanded and produced; and changes in levels of mobility and comfort that people either have or aspire to.

■ The ratio of energy use to GDP provides neither an accurate measure of efficiency in the use of energy nor a gauge for how efficiency improves and deteriorates. Using the apparently simple ratio between energy demand and GDP produces misleading comparisons between countries. It obscures many factors which do differentiate countries and which explain why CO_2 emissions differ among them.

■ No one has repealed the loose correlation between energy use and GDP, but it has lost much of its strategic value for policy-making, for short-term forecasting and for understanding how energy demand may develop in the long term. Policy makers still need aggregated information lest they lose their way in a labyrinth of frustrating detail, but the nature of that information has changed radically. Truly useful knowledge now builds the aggregates from the details. It goes to the heart of myriad and complex decision-making processes for energy use, to the actual human beings who make the crucial energy choices. With information of this sort, analysts and legislators can know how and why energy use changes or remains stable; they can begin to frame policy strategies and programmes that meet today's concerns.

■ The most acute of those concerns centre on the environmental problems associated with the production, transformation, distribution and consumption of energy; they focus on fossil fuels, whose combustion leads to CO_2 emissions. As energy demand continues to grow, experts and authorities seek strategies to moderate that growth, particularly the demand for hydrocarbons. Those strategies will succeed if they accurately distinguish among the many human and social sources of demand, tailoring energy efficiency to reality rather than to stylised, aggregate measures of reality which conceal more than they reveal. Good strategies will recognise where progress occurs and where situations run out of control. They will take advantage of favourable possibilities – and such possibilities do exist – where strong environmental gains can arise with a high probability that the policy invoked can gain political acceptance. They will grow out of solid analysis of potential benefits, as well as environmental, economic and political costs. The strategies will differ. As strategies seek greater gains, they will become much more attuned to their "realities" and will differ across countries much more than in the past. And the nature of the information underlying and informing them can itself make international co-operation easier, more amenable to monitoring and less contentious.

Why This Book?

■ This small volume distills for general readers the essential messages of the IEA's recently published *Indicators of Energy Use and Efficiency*. It outlines how a system of relatively straightforward indicators of energy end-use has developed. It demonstrates how they can effectively contribute to understanding the complex fabric of energy demand in industrialised countries. It extends the use of indicators from energy to carbon emissions, using a few simple assumptions.

■ Although the IEA Member countries achieved a remarkable reduction in the amount of carbon they released for a unit of production or consumption between 1973 and 1993, the forces underlying that reduction have weakened significantly. Carbon emissions are now rising, not falling. In this context, no strategy for future restraint can succeed unless policy makers take into account the key tendencies revealed by indicators such as those developed in this study. Using the indicators, they can spot underlying trends that raise or lower energy use and emissions. They can understand some of the links between those trends and the forces that drive them and they can evaluate differences in trends among countries. An understanding of all of these elements in the present energy situation is crucial to the design of effective carbon restraint policies for the future.

International Energy Agency Energy End-Use Indicators Project

The aim is to take existing national data and push them further so they can be used in an international context, particularly for energy and climate policy-making. In collaboration with the IEA, the data have been analysed in a series of studies carried out for national governments at the US Lawrence Berkeley National Laboratory, the French Environment and Energy Efficiency Agency and the Dutch University of Utrecht. Analytical reviews, workshops and the development of expert information networks examine how energy use interacts with human activity and which indicators best describe these interactions.

This volume summarises existing indicators, drawing on examples from IEA Member countries as analysed by the IEA secretariat and its collaborators. The recently published, **Indicators of Energy Use and Efficiency***, illuminates on a more detailed basis the impacts of the millions of energy decisions individuals make on energy consumption and* CO_2 *emissions. A forthcoming volume will expand significantly on the details of the methodology and results, with particular emphasis on efficiency of energy use.*

Recognising the significance of a disaggregated approach to understanding energy end-use trends, the project looks to extend its application to a broader range of countries. A number of IEA Member countries, including Australia, Denmark, Canada, the Netherlands and New Zealand, have recently requested the IEA to extend the indicators work to include these countries in greater depth.

■ The "indicators" system builds on data that are often freely available in developed countries. Equally often, however, such data are badly organised and little exploited by the many ministries and agencies which collect them. Indeed, the development and use of energy indicators can help build ties among public and private decision makers whose actions affect roads, vehicles, homes and buildings, factories and farms.

■ Further development of the indicators system has a low marginal cost – a point which fiscally-pressed governments will appreciate. Building the system depends on a bottom-up approach that links energy uses to economic activity and energy intensities – energy use per unit of output – to explain differences in total energy use per capita or per unit of output. Indicators at levels of disaggregation that really tell a story will nevertheless support aggregation techniques (which are rather more complex than just "adding them up") that do not lose sight of the plot. Thus, policy makers can obtain simple, useful summary measures on which to base programmes and monitor them. They can also develop the means for fair, accurate international comparisons to support co-operative action and learning.

■ Fundamental social forces and policies quite unconnected to energy use affect housing, industrial structure, urban design and – vitally – transport infrastructure. These forces vary greatly among countries and societies and they have profound impacts on energy consumption. Sometimes their effects reinforce the energy-boosting influence of rising incomes; sometimes they augment tendencies toward moderation; sometimes, they may lead to the saturation of energy markets as human needs are perceived to be met (no household needs six refrigerators or nine automobiles). The indicators aid in the discovery of how all these forces affect the key links between energy use and human activities.

They can point towards policies or policy modifications which will build energy efficiency into the very structure of life, the human and institutional behaviour behind all energy consumption.

■ The indicators reveal many new insights, some of which make sense intuitively, while some come as surprises. The following list offers just a flavour:

- Energy efficiencies increased significantly for many years before the first oil shock of 1973, improving fastest in periods of rising energy prices or rapid economic growth.
- Efficiency accounts for only part of the story, however. Structural changes brought significant shifts in energy use – increases more often than decreases. Similar trends prevailed in the services which energy provides. More efficient engines and designs have made automobiles today larger and more powerful than 20 years ago, but such improvements did not always lead to large reductions in fuel use per kilometre. Between 1973 and the early 1990s, the ratio of fuel consumption to distance driven in autos dropped by more than 30 per cent in North America (where successful, active policies were in force) but by only 10 to 15 per cent in Europe and Japan. Meanwhile, the indicators continued to show more and more kilometres driven, everywhere.
- In the same years, the energy intensity of home heating (the ratio of heat used to home area) plummeted by from 20 to 50 per cent in IEA countries.
- The ratio of fuel use to output in manufacturing plunged by 20 to 50 per cent, a remarkable drop considering that the same ratio for electricity held constant or increased slightly. CO_2 emissions from manufacturing fell relative to output in every country studied, and

Figure 1

Impact of Changes in Energy Intensities on Primary Energy Use*

1970-1994

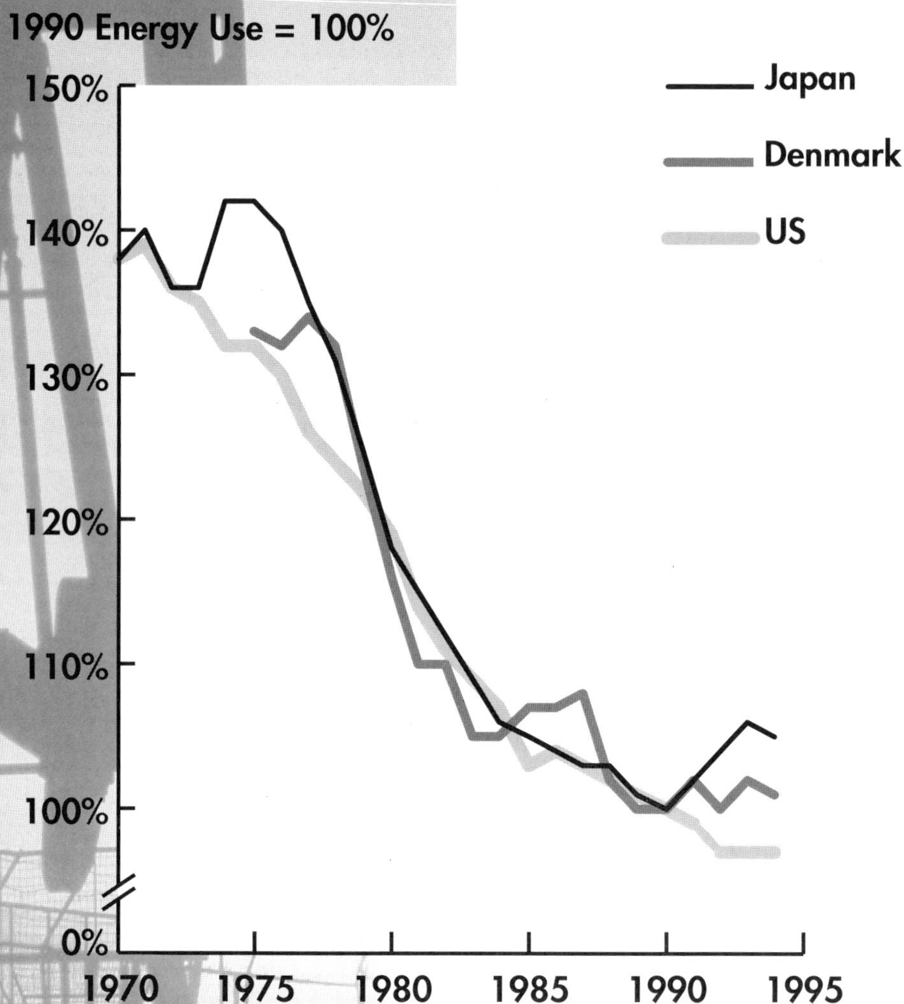

** All sectors, holding structure constant.*

Source: IEA, *Indicators of Energy Use and Efficiency*, (OECD, Paris) 1997.

absolutely in most of them. Manufacturing saw by far the greatest absolute reduction in emissions of all the sectors IEA examined; the indicators explain how and why this happened.

- One danger signal: energy efficiencies generally have improved less rapidly in the 1990's than previously.

■ Figure 1 offers a good summary of the patterns of energy saving over two decades in a sample of IEA economies, with a key indicator developed for the project; for each country it combines in a novel way changes in some 32 energy intensities (efficiency measures) to construct an overall performance figure. It shows a range of quite impressive energy savings which slowed markedly in the early 1990s.

Indicators and CO_2 Emissions: A Way Forward

■ Great differences in the structures of national economies plague international negotiations to set targets for greenhouse gas emissions. Some countries emit more CO_2 because they have cold weather and must heat their homes more. Other emit CO_2 because they have a great deal of heavy industry. The first step to effective international co-operation involves knowing what these differences imply and how they may affect policy. The indicators approach permits the development of "consequential" indicators that link energy use and economic activity to CO_2 emissions. They have many uses.

■ Policy makers in individual countries can use consequential indicators to discover where their principal emissions problems reside and where policies should focus to make emissions restraints effective. In the IEA countries, more efficient energy use and, in most countries, primary fuel switching, cut CO_2 emissions markedly during the 1970's and 1980's. In many cases, such cuts offset the effects of growth and structural change that would otherwise have raised energy demand. The ratios of emissions to activities for most sectors fell by from 20 to 40 per cent, led by manufacturing and space heating (the heating of homes and buildings); but in most countries cars and trucks in the early 1990's belched out higher emissions per unit of output than in the early 1970's, and higher shares of total emissions (Figure 2).

■ Energy end-use indicators break down the differences among countries to suggest how and why some countries find it easier than others to restrain or reduce emissions. Clearly, an understanding of what encourages or inhibits CO_2 mitigation from one country to the next helps to reduce contention and promote better negotiations. With it, decision makers and negotiators can more readily find policies, experiences or technologies which are amenable to transfer, and appreciate how hard it is to transfer others.

Questions, Questions

■ Let us set out now to explore in greater detail how the indicators system works and what it already has revealed about energy use and greenhouse gas emissions in the IEA countries. The task poses a number of key questions.

■ What precise links connect the activity of individuals and groups with energy use, economic life and the environment? Why is an understanding of energy use patterns over a relatively long period in the past essential for projecting them (and changes in them) into the near future or even decades ahead? Why do international differences in these patterns arise and what do they signify? What can indicators

Figure 2

Carbon Emissions per Capita Relative to GDP

1972/73 and 1994

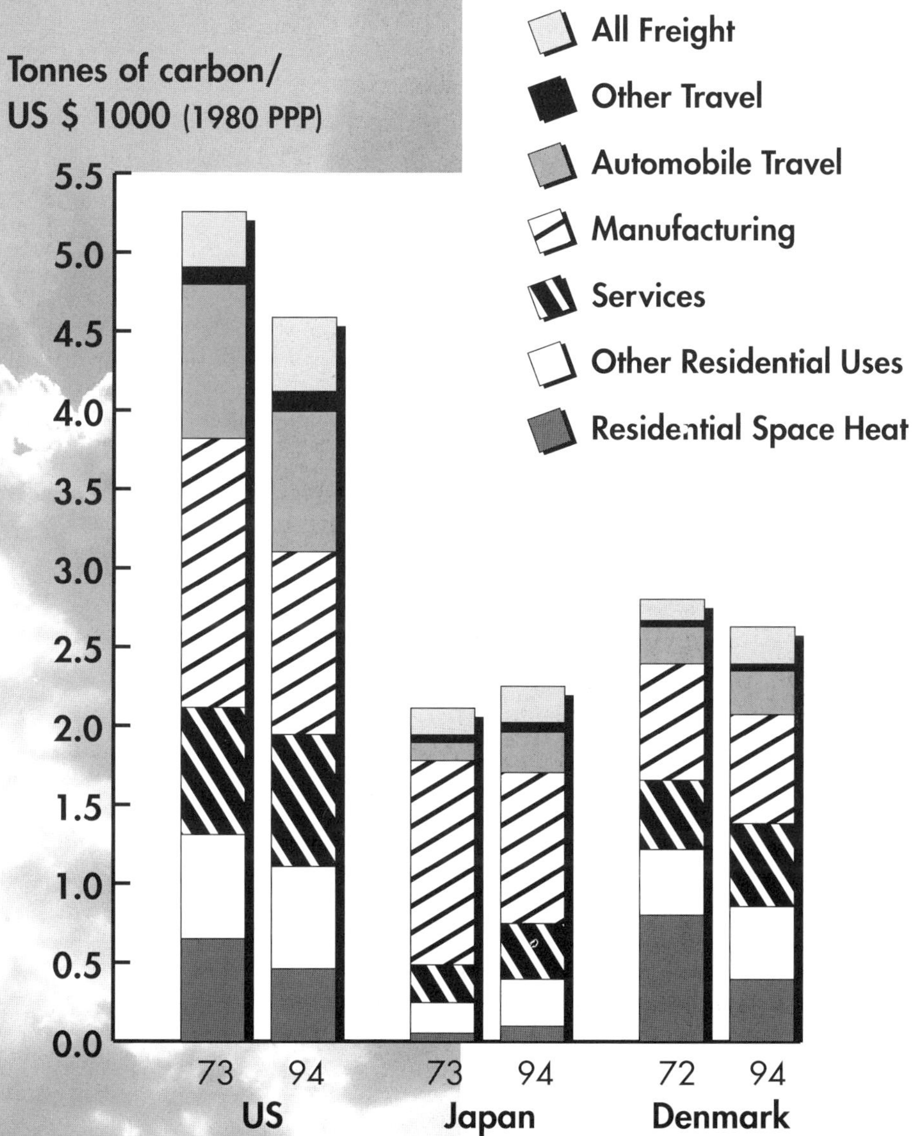

Source: IEA, *Indicators of Energy Use and Efficiency*, (OECD, Paris) 1997.

contribute to the knowledge essential for making good climate policy? What aspects of energy consumption and emissions resist change or, conversely, are susceptible to change for the better? What policies have worked in the past, and why? What will work in the future?

■ To answer these questions, energy indicators must uncover good information on several fronts simultaneously. They need to look at how energy use occurs within the mix of human activities it supports; the quality of its use in the characteristics of the devices through which it passes as well as the utility and comfort its consumption yields; how much energy gets used for given tasks (energy intensities compared to output); its efficiency in the technical sense (which stresses conservation of inputs); and, finally, how different fuels, and therefore different emissions, relate to each energy use or intensity.

Data
Sources,
Building Blocks,
and
Key Terms

THE INDICATORS

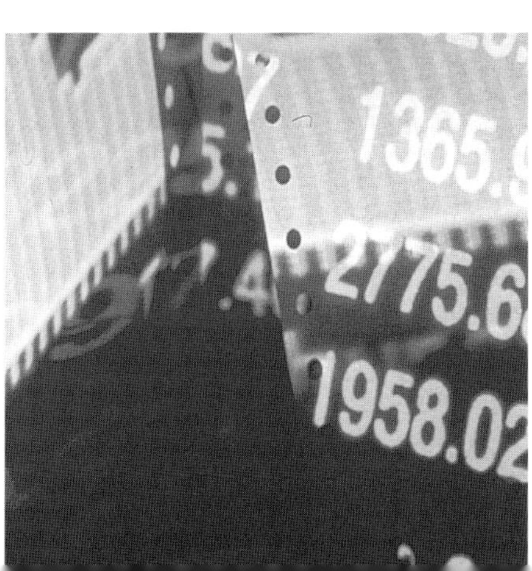

The Big Picture *versus* Building Blocks

■ Energy indicators relate energy consumption to measures of output in ways that provide fine detail at the level of discrete activities. They may also be used to build broader aggregations. The aggregate indicators – built with a bottom-up methodology and not to be confused with misleading numbers calculated from aggregate data – give a global perspective on how broad energy categories have evolved. They show, for example, an important shift in the 1970's and 1980's from direct use of fossil fuels to electricity and the consequent development of a strong link between GDP growth and electricity consumption. As incomes grow, consumers seeking more comfort and firms investing to stay competitive use more electric power. The broad indicators also reveal how the two oil crises deeply affected the use of fossil fuels through price effects which resulted in much fuel saving. And they confirm the importance of GDP as one driving factor in energy use.

■ Policy makers, of course, like aggregate measures for their ease of application. Correctly constructed, such indicators can provide much insight. Yet they lack the precise detail that reveals the effects of many of the other key drivers of energy end-use – the mix of needs to be satisfied, technological changes and human behaviour itself. Because these forces can change energy intensities and usage patterns, analysts want increasingly to see where consumption actually takes place (driving autos, heating, making toast, forging steel), how efficiently it does so, and why. Disaggregated indicators serve this end. At the same time, they feed the aggregation process to yield measures of how the hidden driving forces affect total energy use. Combined with indicators of the final fuel mix for each activity and of the mix in the power sector, the set becomes sufficiently complete to measure which factors boost or restrain CO_2 emissions. The indicators, in short, constitute a *system* and not merely a butterfly collection of data series.

An Analogy

The stock market provides an analogy. Stock market indices, like the broad energy indicators, represent both simple and complex characteristics of a system in an uncomplicated way that remains faithful to the key trends within it. They tell a true story. Yet no investor tracking an individual stock or group of stocks can rely on an index alone. He or she will seek complementary information on individual economic sectors and on the performance of stocks within them, and study the forces which drive that performance. Similarly, the energy policy maker can learn much from aggregate indicators, but targeted, potentially effective emissions-reduction policies (selecting the "high performers") depends heavily on getting to the right level of detail and identifying the driving forces or behaviour which policy is meant to affect.

An Example: Automobile Fuel Use

■ Consider a disaggregated indicator of energy use – automobile fuel economy or intensity. All drivers, especially buyers of new cars, know about fuel intensity, usually expressed as litres of fuel/100 km of driving, or fuel economy, expressed as miles per gallon. Most drivers understand that test results for a new car usually overstate the fuel economy it will actually achieve. They also know that traffic conditions and driving behaviour (speed, acceleration, use of the car when the motor is cold) both affect fuel economy. Total fuel consumption is defined as fuel economy times distance travelled. One can measure actual fuel economy to determine whether efforts to reduce fuel consumption (without reducing distance driven) have paid off. Depending on the results, a person might consider buying a more fuel-efficient vehicle sooner than otherwise foreseen, or taking the bus, the train or an air plane more often. He or she is likely to choose one of the latter options if transport and infrastructure policies offer the right incentives and meet the perceived requirements for business, pleasure and convenience at equivalent cost. Conversely, if the person gets a raise at work, the result might include purchasing a larger, more powerful automobile capable of covering greater distances on the roads because his perceived needs will have changed as a result of the raise.

■ Note that this one example, at the level of an individual, contains all the elements for which the policy maker requires measurement in an indicators system: the objectives of personal travel as they are conditioned by income and a complex mixture of social and cultural forces; energy intensity (distance driven); energy efficiency (the technology built into the automobile); the mix of travel options from which the person can make choices; and energy prices, which affect choices as well. They all combine in complex ways to yield one individual's automobile fuel consumption. A change in any of them, which could arise from a new policy-induced incentive or disincentive or happen independently, will change the result.

■ Moving to higher aggregation levels, analysts can measure the fuel economy (or fuel intensity) of each vehicle type, calculate distance travelled for the same sample of vehicles and fuels, then multiply the two to get a reasonably accurate picture of total fuel use for each kind of vehicle or transport mode. Some IEA Member countries' transport authorities painstakingly estimate total fuel use in this way. Indeed, most IEA countries have at one time or another had goals or policies to lower fuel intensities; so this indicator has become important for energy policy and not simply an interesting measure of fuel economy.

■ Multiplying kilometres per vehicle and litres per kilometre by the total number of vehicles, scaled to the population, gives total fuel use by a bottom-up method, showing how indicators often function simply to help us understand the components of energy use. How each of them has changed will keenly interest national authorities, transport executives and international environmental negotiators as well.

■ Indicators typically come in pairs or trios which can be multiplied together to give more aggregate indicators. Activities for which energy is used may be measured as area heated, tonnes of steel produced, output in monetary terms, etc. Comparing these indicators requires some art but, clearly, aggregate indicators of energy use can be built up.

The Energy Indicators Pyramid

■ The pyramid diagram shows schematically how the most detailed data and indicators can combine. The bottom of the pyramid represents extreme disaggregation; the top displays an aggregated result. The pyramid thus portrays a hierarchy of energy indicators. It also shows how the bottom-up approach, which builds the aggregates, and the top-down approach which starts with them, come together conceptually. The higher-level aggregates match, but the indicators method is the result of much deeper disaggregation and uncovers far more information.

Energy Indicators Pyramid

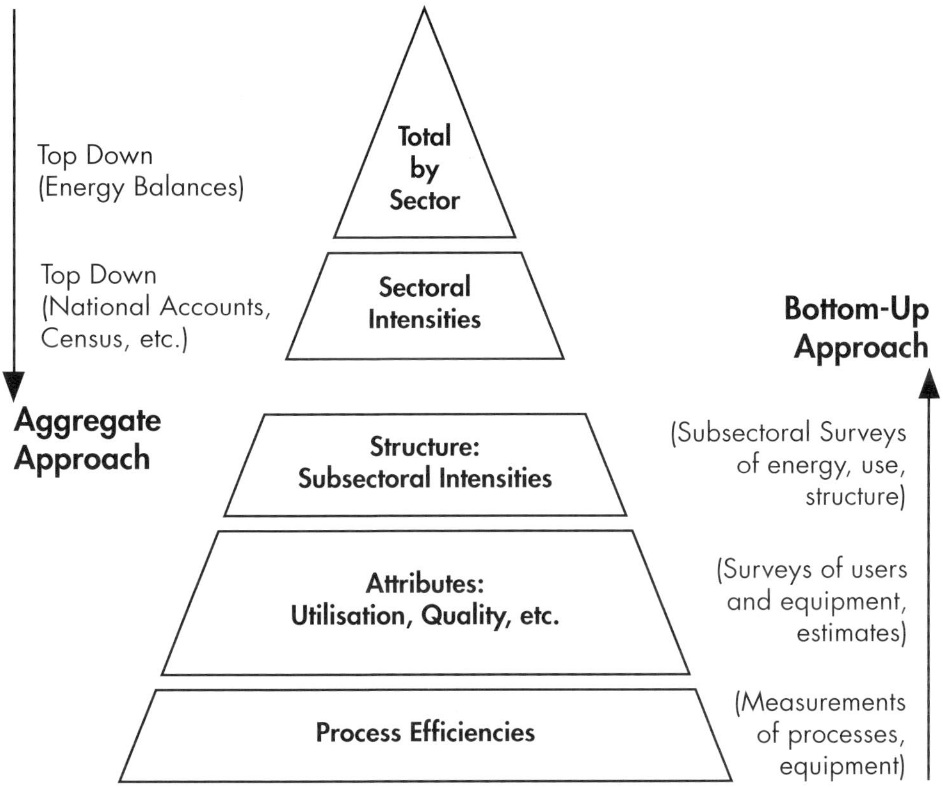

Source: After K. Blok, University of Utrecht.

■ The pyramid could have presented a different set of labels for each measurement task. The top element, for example, might represent the ratio of energy use to GDP. The second row of elements might contain the energy intensity of each major sector. The third row might depict the subsectors or end-uses that make up each sector, while the lowest row (which is not necessarily the final row) might carry even more detail. Structural variables are factored into each level of energy intensities and indicate how to weight them to move up to more aggregate levels. Obviously each descent in the pyramid requires more data and more complex analysis to re-aggregate back to a higher level; but each descent also brings the analyst closer to the engineer's concept of energy efficiency, defined for a specific technology, process, or task.

■ Exactly how far to probe depends on what policy makers need. In the automobile example, if the aim is to make each kind of car more efficient, then individual car types (and their uses) need study. If policy aims only at reducing average fuel use per kilometre driven, then deep disaggregation becomes unnecessary unless the analyst needs to know which kinds of cars were driven the most. Most policies on car travel pursue both objectives. Keep in mind, however, that increasing the energy efficiencies of cars may not translate directly into lower intensities of driving because traffic conditions and driving behaviour also affect fuel economy. Moreover, some increased efficiency may translate into more power for a given fuel use, as is the case for most cars sold in the 1980's and 1990's in North America, Japan, and Europe. Finally, reduced fuel intensity tends to stimulate a small increase in driving. Thus the pyramid reminds analysts that they must always study efficiency, intensity, activity, and actual energy use together.

What Is A "Sector"?

The definitional scheme used here and in the IEA's Indicators of Energy Use and Efficiency *departs from that used in the national accounts because the objective of identifying major energy users demands it. Yet the word "sector" remains in use and it takes some liberties with the national accounts concepts. "Consumption" means using energy and nothing more. The "sectors" covered here include (1)* Households *(energy use in homes), (2) personal* Travel*, (3)* Freight*, (4)* Manufacturing*, and (5)* Services*, which consist largely of heating and lighting buildings not occupied by households, and running the computers therein. None of the "sectors" as defined here, except possibly manufacturing, conforms with the national accounts definitions. The sums of sectoral energy expenditures thus cannot be compared with GDP by sector from the national accounts except after considerable modification of the latter to conform with the energy-related definitions.*

■ This hierarchy permits one to relate very small elements from below, (often the direct results of earlier policies, technological progress, structural reform or behavioural change) with higher-order, more aggregate quantities. One can then show how changes in the former relate to changes in the latter. With this hierarchy one can better explain aggregate changes in energy use in terms of their components, and more carefully chose the depth of analysis required. That choice depends on the questions that need asking to make good policy.

Indicators Come in Different Flavours

■ *Descriptive* indicators show energy use and its changes by sector and, at deeper levels of detail, by its end-use. *Structural* indicators show how the activities or products associated with energy use, such as steel production, driving vehicles or heating home areas, have evolved. Combining these indicators with energy uses gives *energy intensities* (*i.e.* energy use per unit of output) which relate to efficiency. Somewhat more effort can yield actual *efficiency* measures as well. All these measures lend themselves to international comparisons. Various techniques for index *decomposition* help establish how these same elements have affected total energy use and CO_2 emissions over time. Relating changes in indicators to variations in income, energy prices, demographic structure and other macroeconomic elements or policies produces *causal* indicators that show which forces stimulate energy use, which restrain it and which evoke more efficiency. Such insights play key roles in sound policy formulation. By adding the primary fuel mix to each end-use, one can relate changes in structure and efficiency, as well as the fuel mix itself, to changes in carbon emissions; this produces one kind of *consequential* indicator.

■ The indicators themselves cover all major sectors of energy end-use. A relatively short list of key uses – car driving, freight haulage, home heating, production in energy-intensive industries – can indicate the major sources of changes in consumption. With the indicators, analysts can evaluate past efforts to affect energy use and design new policies for carbon restraint, always in the context of a larger set of driving forces.

Key Terms

Delivered energy or final consumption:
Energy supplied, for example, to a building, factory or fuel tank and converted ultimately to heat, light, motion or other energy services. Transformation and distribution losses are not included.

Primary energy:
Delivered energy plus losses incurred in converting energy resources into purchased heat and electricity.

Useful energy:
Delivered energy minus losses assumed to occur in boilers, furnaces, water heaters and other equipment in buildings; used for estimates of heat provided in space and water heating and cooking.

Energy intensity:
Energy "consumed" per unit of activity or output.

Activity or output:
Basic unit of accounting for which energy is used, e.g. in space heating, it is the area heated; in manufacturing, it is the output in tonnes of steel or number of widgets.

Structure or subsectoral activity:
Refers to the modal mix (trucks, rail, ships) in travel, energy end-uses in households, and manufacturing output by branch.

Energy services:
Implies actual services for which energy is used: heating a given amount of space to a standard temperature for a period of time, etc.

Indicators of Energy Use in the Residential and Manufacturing Sectors

HOUSES, FACTORIES

3

Some Real Results

■ The IEA's recently published *Indicators of Energy Use and Efficiency* reports exhaustively on the state of the art of developing indicator systems. It conveys the many facts and insights which this work reveals about energy use in IEA countries. This chapter and the next present the key results by sector; Chapter Five integrates them.

■ The table on the following page contains a number of surprises. It presents estimates of energy-related carbon emissions by country and sector for five IEA countries in 1973 and 1994. The first surprise is the relatively small reductions (in some, cases, even increases) in absolute levels of emissions over roughly two decades. The trend extends to most of these countries, including those which, taken together, dominate IEA totals. Emissions per capita, on the other hand, dropped sharply, and emissions per unit of GDP plunged in every country and sector except transport. Impressive energy savings, bolstered by relatively modest official energy-saving programmes, did not, in most of the IEA countries, offset economic growth and other factors pushing energy use upward.

■ Stronger economic activity and some structural changes – especially in the travel, freight and household sectors – raised emissions in all countries, while other structural factors worked the opposite way in six countries. Declining energy intensities across all sectors operated as the main factor restraining emissions. A long-term trend away from carbon-intensive fuels in the overall fuel mix also helped. The shift was dramatic in countries adopting nuclear power (France, Finland and Sweden). Switches to natural gas now under way in Denmark, Germany and the United Kingdom push this trend further.

Estimated Carbon Emissions from Energy End-Uses in Selected Countries, 1973 and 1994

Countries and Sectors[1]		Levels of Emissions					
		Absolute (Millions of tonnes)		Per Capita (Tonnes)		Per Unit of GDP[2] (Grammes)	
		1973	1994	1973	1994	1973	1994
United States	Sectoral Totals	1 113.9	1 196.0	5.3	4.6	473	309
	Manufacturing	361.1	302.1	1.7	1.2	154	78
	Households	277.8	288.6	1.3	1.1	118	75
	Travel	230.1	265.0	1.1	1.0	97	68
	Services	170.7	218.3	0.8	0.8	73	57
	Freight	74.2	121.9	0.4	0.4	31	31
Japan	Sectoral Totals	229.6	281.3	2.1	2.3	317	186
	Manufacturing	140.5	119.5	1.3	1.0	194	79
	Households	26.8	49.4	0.3	0.4	37	33
	Services	26.0	44.0	0.2	0.4	36	29
	Travel	17.7	39.8	0.2	0.3	25	26
	Freight	18.6	28.6	0.2	0.2	25	19
W. Germany[3]	Sectoral Totals	187.2	171.2	3.0	2.6	345	207
	Manufacturing	82.2	56.7	1.3	0.8	152	68
	Households	50.5	51.5	0.8	0.8	93	63
	Travel	17.0	27.5	0.3	0.4	31	53
	Services	29.9	26.4	0.5	0.4	60	45
	Freight	7.6	9.1	0.1	0.2	21	21
United Kingdom	Sectoral Totals	167.6	134.5	3.0	2.3	451	275
	Households	50.2	41.6	0.9	0.7	135	85
	Manufacturing	72.4	38.8	1.3	0.7	195	79
	Travel	14.9	22.0	0.3	0.4	40	45
	Services	22.3	21.9	0.4	0.4	60	45
	Freight	7.8	10.3	0.1	0.2	21	21
Denmark[4]	Sectoral Totals	14.0	13.7	2.8	2.6	423	265
	Households	6.1	4.5	1.2	0.9	184	86
	Manufacturing	3.7	3.6	0.7	0.7	112	70
	Services	2.2	2.7	0.4	0.5	66	53
	Travel	1.4	1.7	0.3	0.3	41	33
	Freight	0.7	1.2	0.1	0.2	20	23

Notes: Includes emissions from electricity and district heating allocated in proportion to their end-uses. **1.** Sectors are ranked for each country in declining order of contributions to the total absolute level of carbon emissions of that country in 1994. **2.** GDP converted to 1980 US dollars from real national currency at purchasing power parity. **3.** Final year is 1993 for western Germany. **4.** Initial year is 1972 for Denmark.

Sources: Carbon emissions from Lawrence Berkeley National Laboratory (LBNL) of US Dept. of Energy and IEA, *Indicators of Energy Use and Efficiency*, (OECD, Paris) 1997. Population and GDP from OECD *National Accounts* for the relevant years.

Scope of the Project

Indicators of Energy Use and Efficiency contains material from about half of the 24 IEA countries accounting for almost 90 per cent of primary energy consumption in the IEA area. The countries include the "EUR Four" – France, Germany (the western portion only), Italy and the United Kingdom, the Nordic countries, plus the United States and Japan, with data and indicators from Australia, Canada and the Netherlands introduced where available.

The large mass of data contained in the book came "off the shelf." This demonstrates what already is available if one uses it correctly. Primary data came from governments; secondary sources included research establishments sponsored by governments and the European Union.

Star Sectors

■ Manufacturing and households turned in the best performances, cutting their energy intensities deeply. Factories achieved bigger cuts in emissions over twenty years than did any other sector, notwithstanding a probably temporary lull in the last few years. Higher fuel prices, better technology and strong international competition propelled these changes. Higher prices also prodded households to great savings, helped by rising efficiency standards and official programmes where they existed. Energy saving is still occurring in both sectors, where new equipment continues to have less energy intensity than the machines that it replaces. Both sectors show basic trends which good policies could accelerate.

■ Households: John Q. Public Does His Part

■ The residential sector accounts for 15 per cent to 25 per cent of primary energy in IEA countries. Most Member countries gave more policy attention to households in the 1970's and 1980's than to any other sector. That attention continues. What have been the results?

■ With energy indicators, authorities can see how significant changes in the intensity of heating and new appliances, together with saturation in equipment ownership, have reduced per capita household energy use. The indicators also show that energy use per unit of household expenditure has diminished. Without benefit of the indicators, policy makers cannot describe how in past decades activity, structure, and intensity have affected the overall ratio of household energy use to GDP or income. So they cannot use such a ratio as an indicator of the efficiency of household energy use in the future.

■ Two opposing trends marked the 1970s and 1980s. Expanding consumption, driven by higher incomes, might have been expected to raise per capita household energy use by some 30 per cent to 70 per cent in the countries studied. Yet the actual figures increased by 30 per cent at most; they actually fell in some countries, most notably Denmark. Except in Japan, changes in per capita household energy use between 1973 and 1992 were small or even negative. Given the small size of Japanese homes and the sparsity of appliances in the 1970's, the slow increase is understandable; it is clarified by use of the indicators.

■ Some of the energy savings gain occurred as electricity substituted for fossil fuels. Most savings, however, came from more efficient energy use. Reduced energy intensity in the heating of houses and buildings formed the principal component, as a result of small reductions in indoor temperatures and significant improvements to the thermal properties of dwellings. Slow but steady replacement also permitted savings of 10 to 15 per cent in the average unit energy consumption (UEC) of electric appliances, despite steady increases in the size and design of the machines. In every country, trends in household energy use after 1979, and in some cases after 1973, differed radically from the profligate patterns prevailing before. Household electricity use deviated from its historical growth pattern in several countries; it swelled with the rise of electric heating in a few countries and slowed relative to income gains in countries which saw significant improvements in electric appliances.

■ A complex set of changing factors affecting the structure, fuel mix, and energy intensity of households underlies the relatively smooth picture of per capita energy use provided by aggregate statistics. Some of the changes resulted in lower energy intensities, but most boosted them. Overall, the per capita measures converged among countries but some key factors pointed the other way. Not every facet of energy use by households can be linked to changes in the aggregate. Demographic trends (income, age distribution, and household composition), dwelling characteristics (size, construction type, vintage), equipment stock and other structural indicators clearly influence how much energy gets consumed for each purpose.

Energy Intensities and Efficiencies

■ Space heating accounts for the largest share of per capita energy delivered to households (58 per cent in the United States and 30 per cent in Japan in 1992, with other countries falling in between). Next come water heating and electrical appliances, followed by cooking and lighting – all of which can use electricity.

■ Household electricity use grew in every country examined, propelled by increases in equipment ownership and, in some cases, by notable new penetration of electric space and water heating. In the United States, for example, the share of electricity in per capita energy delivered to households rose from 11 per cent to 33 per cent between 1973 and 1994; the same figures for the principal Western European countries were 6 and 20 per cent, and those for Japan 10 and 37 per cent.

■ At the same time, significant, sometimes dramatic declines in energy intensities for heating and appliances also took place. With strong expansion both in areas heated and in the number of electric appliances, the small increases in energy use, and even some declines, strongly suggest increased efficiency. Indicators can decompose the overall changes into components related to increased equipment ownership (a structural factor) and lower energy intensities.

■ Home-heating intensity (figure 3), the most readily obtainable aggregate indicator, also serves as the single most important measure of energy intensity for the residential sector. Government policy has aimed to reduce the energy used in space heating, with evident success in most countries. In 1973, differences in heating practices, combustion efficiencies and heat losses in buildings caused considerable divergence in intensities among countries. Twenty years later, intensities had declined and converged remarkably.

■ Behavioural factors – areas heated, heating time and average temperatures – heavily affect the structure of space heating. Significant interaction between efficiency and

Figure 3

Residential Space Heat Intensity

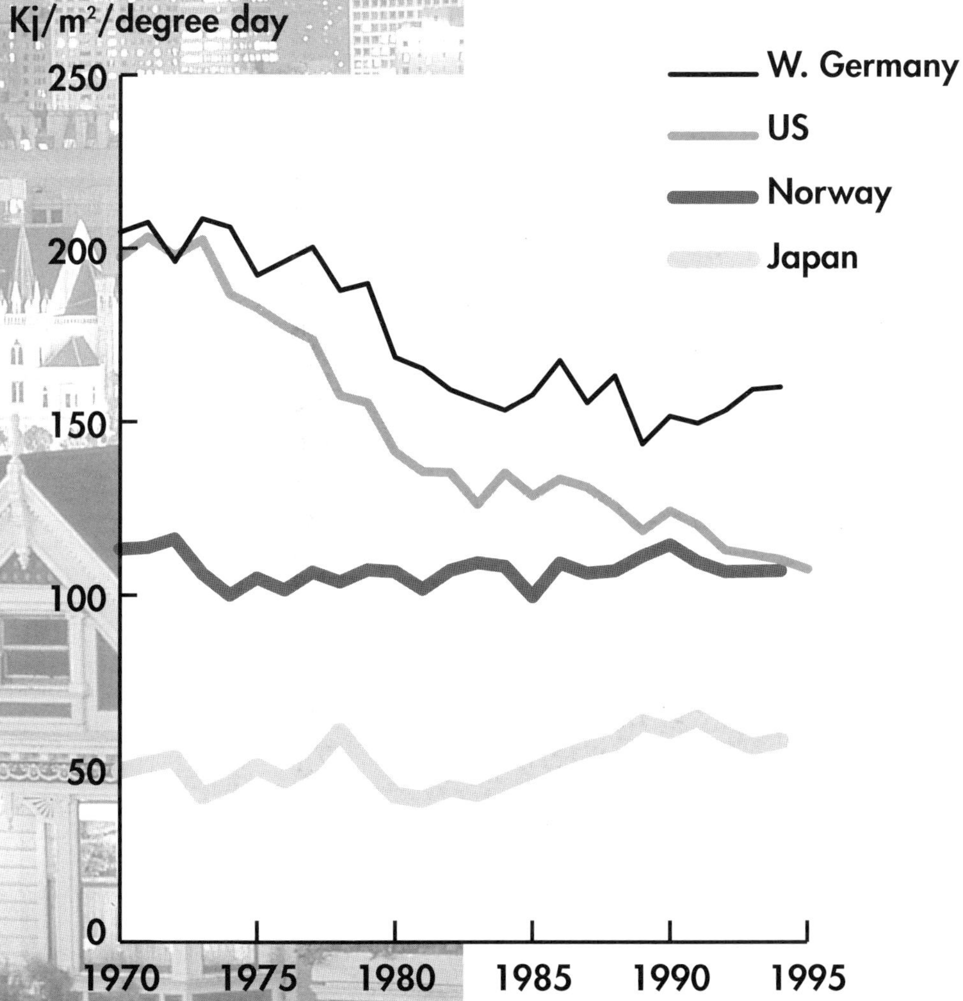

Source: IEA, *Indicators of Energy Use and Efficiency*, (OECD, Paris) 1997.

comfort can occur when the initial comfort level is low. As the insulation of dwellings increased, standards in the UK and Norway improved at about the same rate, leading to a marked improvement in living conditions with almost no increase in space heating intensity. In Japan, where indoor temperatures are kept low by European standards (15-16°C when heat is used) space heating intensities remained low because modest improvements in insulation and increased use of heat pumps contributed to higher comfort with little increase in energy consumption.

■ Most changes in the space heating indicator came from reduced heat losses through improved insulation and windows, but some arose through lowering indoor temperatures, more careful heating practices and improvements in the efficiency of heating equipment. Only a small part of the gain stemmed from fuel switching. This indicator ignores a significant increase in central heating in countries outside the Nordic area; thus the real improvement in those countries was greater than that implied by its statistical decline.

■ Household electricity use has a behavioural element and a technical one. The first involves a vast increase in appliances owned by the average household, and their use in more energy-saving ways. Technical change grows out of an engineering response: more efficient appliances. The average estimated unit electricity consumption (UEC) per appliance and per year dropped for major appliances in all countries. This measure depends both on efficiency and on size and characteristics, e.g. how often clothes and dishes are washed. Nonetheless, the declines in UEC suggest big improvements in new appliances, as well as behavioural changes in the ways appliances are used.

■ The lower average consumption of new appliances curtailed the increase in household electricity consumption that might otherwise have occurred. By 1992, the decrease in average energy consumption by six major appliances (refrigerators, refrigerator/freezers, freezers, washers, dryers, and dishwashers) had reached between 10 per cent in most countries and nearly 18 per cent in first-place Denmark, with the United States in second place. Other nations chalked up smaller declines, in part because the size of some equipment (notably refrigerators and refrigerator/freezers) increased significantly.

Aggregate Indicators

■ No simple aggregate picture of household energy use exists. With so many different energy end uses in the home, changes in aggregates bear little relationship to intensity or efficiency. Modest increases in delivered energy use in most countries hide both a significant decline in intensities and considerable gains in comfort and appliance ownership. UEC provides the closest measure of services delivered. Data on delivered energy most closely measure aggregate consumption; and indicators of primary energy do the same for the energy resources consumed to provide the delivered energy. All three yardsticks can usefully show different effects in the same context.

■ Using indices to decompose the changes apparent in aggregated statistics reveals structural changes underlying them. Most countries except Japan have experienced significant drops in energy intensities, caused mostly by house heating with some declines in water heating and major appliances. But the numbers of appliances and overall heating standards have, in fact, increased. These intensity and structural effects nearly offset each other. Intensities in Denmark and the United States, the countries with the deepest cuts, have not rebounded.

Differences Among Countries

■ After winter climate, the size of homes accounts for the greatest differences in per capita heating energy use, which in turn is the largest component of differences in per capita consumption.

■ Differences appear for other reasons in water heating. The United States and the Nordic countries have high hot-water consumption primarily because nearly all households have large central tanks. Japan uses a lot of hot water for reasons related to family bathing traditions. The remaining European countries have a mix of central and non-central sources that are both associated with lower consumption. The penetration of electric house and water heating ranges from less than 10 per cent of homes to more than 30 per cent (Norway, with 90 per cent represents an extreme).

■ Electric appliances constitute another key area, because the numbers and size of household appliances make them the second largest contributors to differences in consumption. Energy efficiency profiles are changing at varying rates in different countries. US appliances have been less efficient, but are gaining, though the country has a lot of stock to work through the system. Much higher consumption in the United States relates first to the number, size and sophistication of appliances. But it also has to do with efficiencies, with older US refrigerators much larger than those in Europe and probably less efficient. Top-loading US washing machines require more energy than the front loaders more commonly found in Europe. At the margin, however, differences in the efficiency of new appliances have almost converged.

Driving Factors

■ Structural variations eclipse energy intensities as sources of the manifold differences in household energy use. Among them, the economic characteristics of households have as much importance as physical and behavioural characteristics. Income plays a large role. As household wealth rises, the share of expenditure on energy falls, even as households acquire more ways to use it, especially electricity. Increased incomes translate into larger homes, more and larger appliances, and more services such as heating and hot water. The natural result is higher energy use in 1993 than in 1973, despite substantial energy savings for some end-uses. Nevertheless, fuel and electricity use for heating did not rise as fast as growth in home area, central heating, and comfort; a combination of higher efficiency and some demand saturation led to this restraint.

■ Notwithstanding the endless human fascination with new gadgetry as incomes rise, the notion of "saturation" has real meaning for maturing markets which satisfy basic needs like living space, heat, hot water, comfort and ease of household work. This phenomenon reduces the marginal impact of higher incomes on household energy use. It also affects more consumer energy markets now than it did two or three decades ago simply because incomes have risen strongly and most households in developed countries have acquired these amenities. Policy makers can capitalise on this observation.

■ Energy prices are another key driving factor in consumption. Prices have fallen back somewhat in the past few years, but new equipment became significantly – and irreversibly – more efficient during the high-price years from 1973 to 1986. Capital equipment and consumer durables have long lives which perpetuate efficiency gains originally triggered by higher energy prices. Manufacturers of new equipment do not regress or stop advancing technologically when price incentives have died away, especially when they face markets composed of savvy consumers. Some customers really *do* read and act on the efficiency labels which now decorate appliances in the shops.

■ In terms of the household services which energy provides, no single price rules across national markets. Electricity prices showed considerable divergence in 1973 and have evolved somewhat differently since then. They are relatively high in Japan and Germany (and also in Denmark because of high energy taxes), but low in Finland, Sweden and Norway and intermediate elsewhere. In every country except Japan, Denmark and Italy, one energy source remained relatively cheap during the period of high oil prices (1973-86) or else prices for most household energy sources collapsed after 1986. Denmark saw the biggest hike in heating-fuel prices in Europe and high prices persisted through the late 1980's. Not surprisingly, the Danes reduced their household energy intensities more than others did. The Italians and Japanese faced the highest overall prices and now have the lowest consumption, even after adjusting for mild climate. The Japanese and Germans faced relatively expensive energy from 1973 to 1985, but their prices in the early 1990's resembled those of the pre-1973 period. The United States and United Kingdom, blessed with natural gas, also saw prices in the 1990's that were close to or less than those in the early 1970's. The Swedes and Norwegians, with low-priced electricity and a significant quantity of free wood, have reacted least vigorously.

■ Aggressive conservation policies in a few countries also played a role. Aimed mainly at reducing energy use for heat and hot water, they remain in force in Denmark and to a lesser extent in other countries. This may help explain why total household energy use and intensities did not rebound significantly after the crash in oil prices in 1986.

■ Yet the actual impact of such measures resists quantification with the information now available. Few authorities have carefully surveyed the steps taken to save energy in the general population and among those who participated in various energy-saving projects. Even fewer studies have conducted before-and-after measurements to see what saving actually occurred. Certain changes, however, such as strengthened building codes and boiler standards in Europe, or appliance efficiency standards in the United States and Canada, did have measurable effects. The right indicators would have illuminated these unknowns and aided policy formulation immensely.

■ International comparisons of use and price confirm the important role of fuel and electricity prices. Price rises produced most of the changes in energy use as well as efficiency, at least through the late 1980's. The biggest changes in the fuel intensities of house and water heating in most countries occurred rapidly after price increases, whereas residential energy conservation programmes touched only a fraction of households. Most of the savings occurred without help from governments, or reached beyond what various subsidies and programmes tried to achieve. The radical changes in energy intensities after 1979 meant rapid cutbacks in energy services. Gradually, the more permanent effects of technical improvements to building shells, equipment and electric appliances supplemented and replaced them. And, of course the technological improvements remained with us even after oil prices fell.

Figure 4

Carbon Intensity of Residential Space Heat

1970-1992

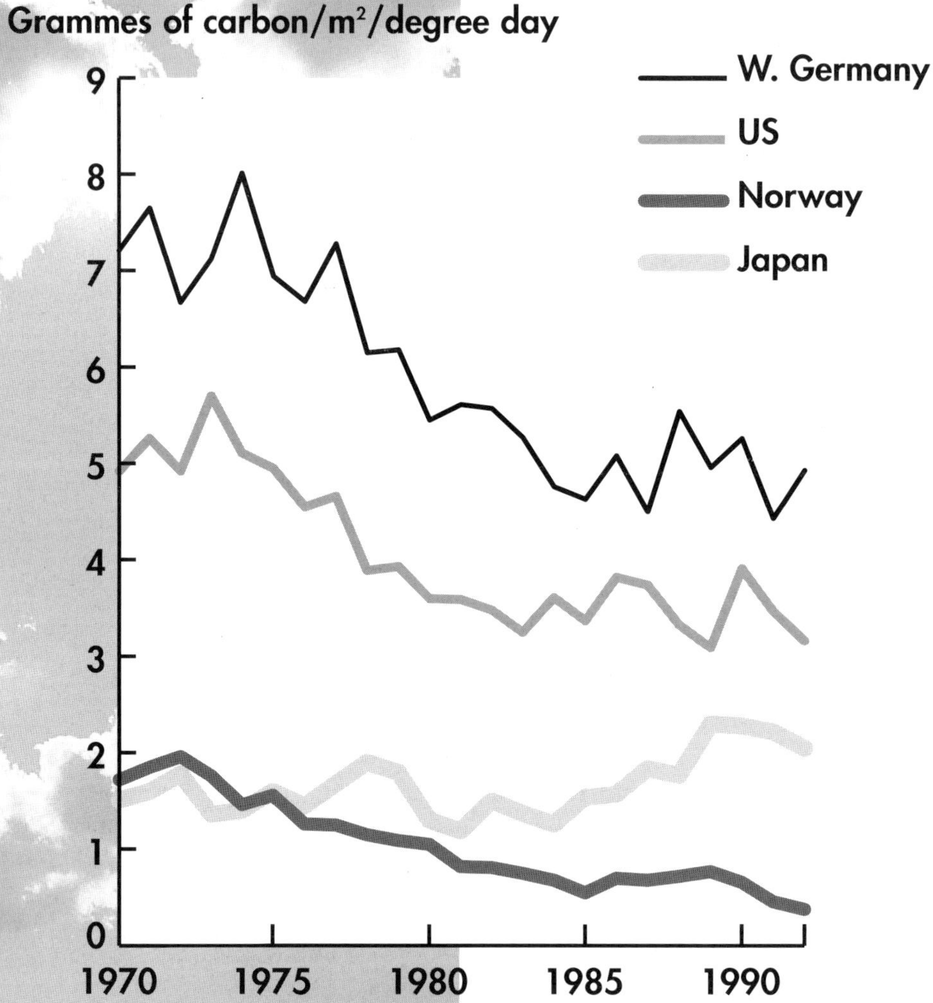

Sources: National studies, household surveys and IPPC co-efficients.

CO_2 Emissions from Household Energy Use

■ The primary fuel mix for households, including both final and utility fuels, became less carbon-intensive, mostly because of changes effected by utility companies. The overall effect cut per capita emissions in the household sector. Although proportions vary by country, roughly half to two-thirds of the reduction arose from lower intensities, the rest from fuel shifts or changes in primary supply. Less carbon-intensive space and water heating dominated, but increased use of electricity boosted emissions in those countries (Denmark, Germany, Italy, the United States and the United Kingdom) where electricity generation is more carbon intensive than the natural gas or oil that it replaces. Increased numbers of appliances boosted emissions in these countries too. The carbon intensity of house heating fell (figure 4), principally because of lower energy intensity and, to a lesser extent, because of changes in the fuel mix. Similar estimates can be made for other end uses.

■ By 1995, structural differences had become much less important than in 1973, but they still accounted for large differences among countries. These structural differences in turn led to differences in CO_2 emissions that neither technology nor public policy can easily erase. The coldest countries had the lowest space heating intensities but they had large dwellings that needed to be kept warm. Denmark had the most carbon-intensive primary fuel mix. American appliances remain more energy intensive than those in Europe (even discounting size differences). Fuel mix also emerges as a key component in the reduction of energy intensities, particularly for electricity, which by the 1990s represented over 40 per cent of the primary energy used by households in IEA countries.

■ Manufacturing

■ Industry in IEA countries – unlike the household sector – has not been subjected to tough governmental energy-saving policies. But it has performed well nonetheless, responding to several intrinsic driving forces. The radical changes in its energy use and intensities over the past few decades require the analyst to understand how the underlying components of demand for manufactured products have changed and how they are linked to output and energy prices.

Structure

■ Manufacturing output, defined as value added or "manufacturing GDP," grew in most countries from the early 1970's to the early 1990's. The increase was concentrated in major branches which show huge variations in energy intensity. A handful of them, such as steel, chemicals and pulp/paper, which account for a small share of total output, nevertheless consume 75 per cent of all energy used in manufacturing. Physical measures, in tonnes of key raw materials, complement the value-added measures of output.

Energy Use and Intensity

■ Policy measures to encourage reduction of energy intensities in manufacturing, which have been rather weak for the most part, have included reporting, voluntary targets, energy management and training, audits, and loans and grants for energy saving projects. In recent years, several countries such as Denmark have introduced carbon taxes, while the Dutch have set tough standards

Figure 5

Impact of Changes in Manufacturing Energy Intensity*

1960-1994

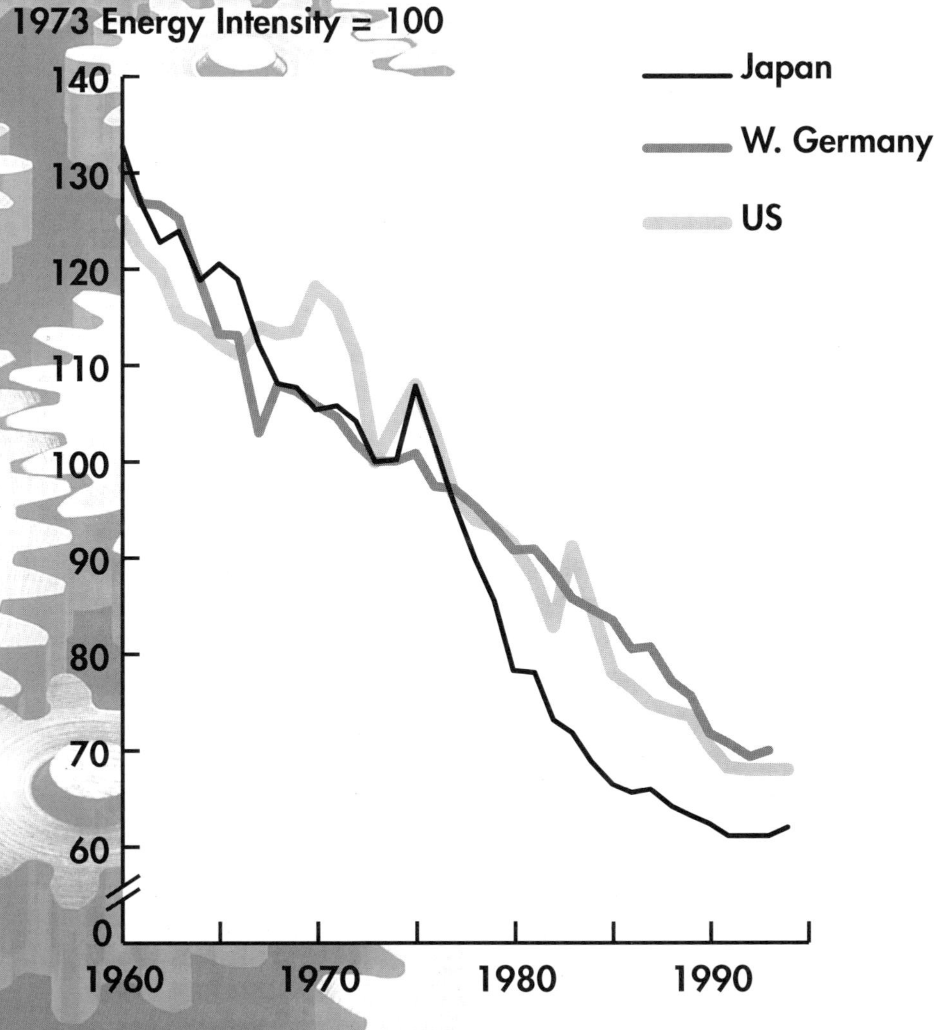

**Weighted index of 7 branches combined with constant 1973 output structure.*

Sources: Lawrence Berkeley National Laboratory, IEA.

Figure 6

Energy Use by Manufacturing Branch

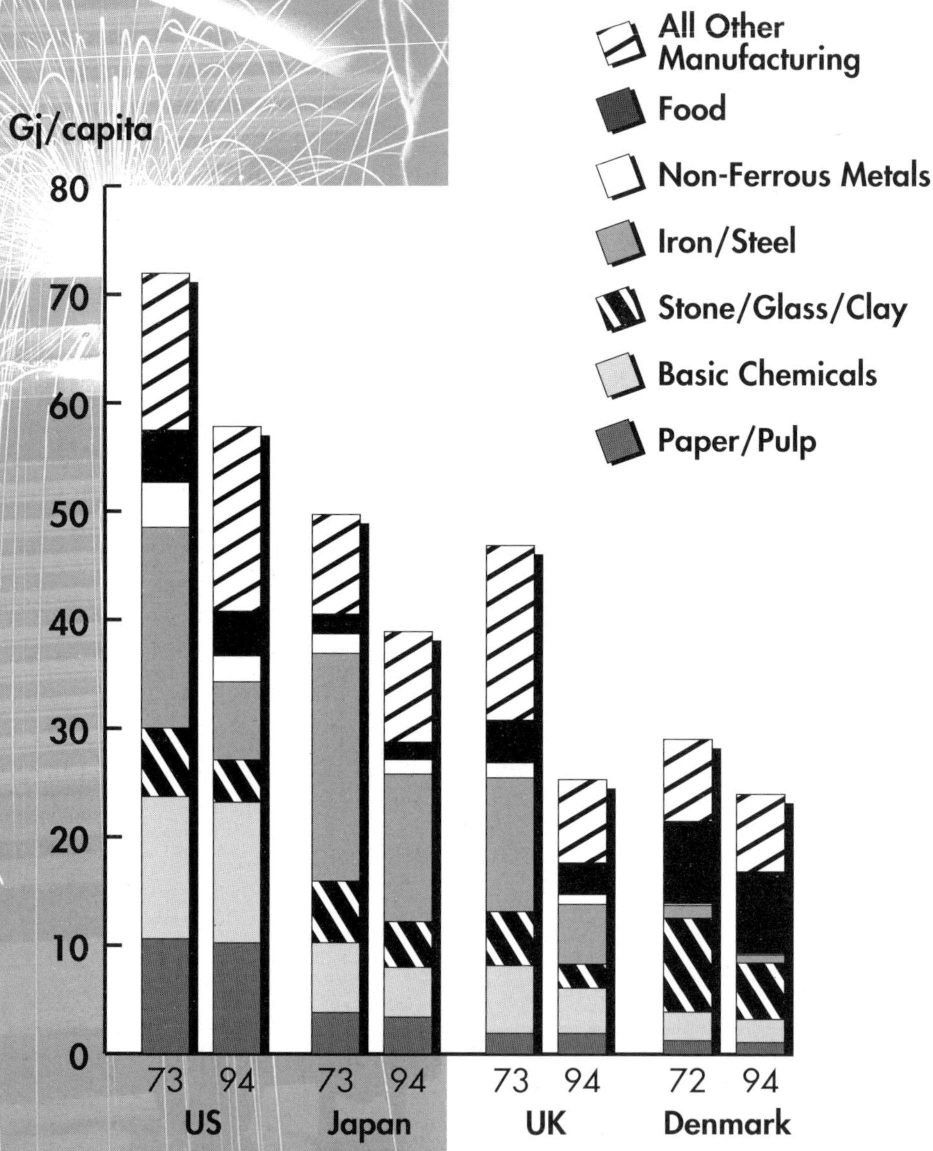

Source: IEA, *Indicators of Energy Use and Efficiency*, (OECD, Paris) 1997.

through binding "voluntary" agreements to reduce CO_2 emissions. To measure the success of such efforts (not an easy matter) requires as a first step an ability to track changes in energy intensities and understand their components. Understanding differences among countries can identify significant energy-saving techniques or policies in one country that might well be applied elsewhere.

■ The mix of major energy-using branches varies by country. For many IEA countries, fairly modest disaggregation captures most of the detail necessary to understand internal developments and make country-to-country comparisons. Ratios of each branch's output to final energy use form the key indicators of subsectoral energy intensity. These ratios declined in almost every branch in every country, a trend well established long before 1973 (figure 5). One result: energy consumption in most branches hardly increased over the 20 years from 1973 to 1993 although output rose significantly (figure 6). More efficient energy use, particularly of fossil fuels, almost offset the growth in output.

■ The efficiency indicator comes from a weighted average of the energy intensities of the six most energy-intensive manufacturing branches, plus "all others," weighted according to their energy intensities in 1973. It too shows steady improvements from long before the first big oil price increase in 1973; technological progress and competition clearly influence energy intensity and consumption independently of prices.

Fuel Choice

■ Fuel mix, an important determinant of pollution and CO_2 emissions, demonstrates a number of key trends:

- A steady move toward natural gas and electricity and away from coal and oil since well before 1973;

- The declining importance of iron, steel and non-metallic minerals from the early 1970's to the mid 1990's, and increased use of biomass in pulp and paper manufacturing;
- Dramatic declines in the share of oil since the early 1970's in all but a few IEA countries; if this did not constitute a direct response to a key objective of energy policy, it certainly was in consonance with it;
- A rising share of electricity in each country, for three reasons: first, and most important, a drop in the ratio of fuel use to output (fuel saving); second, a continuous rise in the penetration of electricity for mechanical processes (lifting, spinning, rolling, etc.); and third, electric-intensive processes that indirectly reduce fuel use (e.g. electro-static paint drying, electro-steel making, electric glass melting).

Aggregate Indicators

■ The global indicator - final energy use per unit of GDP in manufacturing - primarily reflects improved energy efficiency. It fell smartly in the 1970's and 1980's. Disaggregated indicators yield a relatively straightforward explanation for this dramatic drop relative to GDP and for the decline in the sector's share of energy use. First, fuel intensity per unit of output fell in every country, mostly as a result of increased efficiencies. At the same time, electricity intensity dropped very little or increased. Second, shifts in the output mix led to declines of more than 10 per cent in energy use in three countries, less electricity use in two and a big increase in electricity use in one. Consequently, the global decline in manufacturing energy intensity resulted primarily from falling fuel intensities in each major branch. Indicators also show that the so-called "exporting"of industries played only a minor role in the reduction of energy use as compared to manufacturing GDP. The only countries where this happened were Japan, the United States and Germany. The share of manufacturing in GDP fluctuated; only in a few countries did it decline steadily and significantly.

■ A word of caution: only some of the differences among countries arose from differences in energy efficiency. The indicators provide very good measures of changes by country in energy use and efficiency, but to understand differences among countries, the observer must look to their different structures of production. Comparing energy use in manufacturing among countries and branches can be a straightforward matter at an appropriate level of detail, but is potentially misleading at the aggregate level. Many factors besides intensities and prices lead to the significant differences in energy efficiencies which drive the differences in subsectoral energy intensities.

Driving Forces

■ Prices, economic growth rates, technological innovations and capacity utilisation all influence the economic environment that determines manufacturing energy intensities. Energy prices and the rate of economic growth drive energy intensities beyond their long-term downward trends. During brief recessions, declining capacity utilisation can increase energy intensities. Although energy intensities were already falling in the 1970's, higher energy prices after the oil shocks brought additional declines. Most of this happened as energy-saving technology substituted for energy itself. Energy intensities decline with events such as the closure of less efficient plants during recessions and when companies restructure or downsize, actions which tend to raise average energy efficiency. Capital costs –

which were high, particularly in the early 1980s – affect the relative attractiveness of energy-saving investments. The rate of growth in output influences the rate of gradual reinvestment in new technology. Increases in the scale of production also affect energy intensities. The likely lag time between investment decisions and bringing new capacity on stream can create delays between the onset of price increases and a decline in energy intensities.

■ Other factors count too, especially in the long term. In Norway, pulp manufacturing relies on electricity-intensive processes like thermo-mechanical pulping, while in Finland and the United States chemical pulping plays a stronger role. Japan, rebuilding after the war, moved rapidly to more modern-steel making facilities, while the United States and Britain slowly shed older, less efficient technology. Because of intense international competition, all producers have had to cut costs to stay competitive; this means saving energy as well as other resources. With all these forces at work, the energy intensities of different manufacturing branches have to some extent begun to converge, branch by branch, among countries.

CO_2 Emissions from Manufacturing

■ How much can manufacturing reduce its CO_2 emissions? In the early 1990s manufacturing industries held about level with early 1970s values even with increases in output; carbon intensity per unit of output plunged (figure 7). This indicator shows how the ratio of carbon to output, holding all other things being equal, varied because of efficiency improvements (the major factor), changes in final fuel mix, and changes in the fuel mix of utility companies (crucial in

Figure 7

Carbon Intensity of Manufacturing*

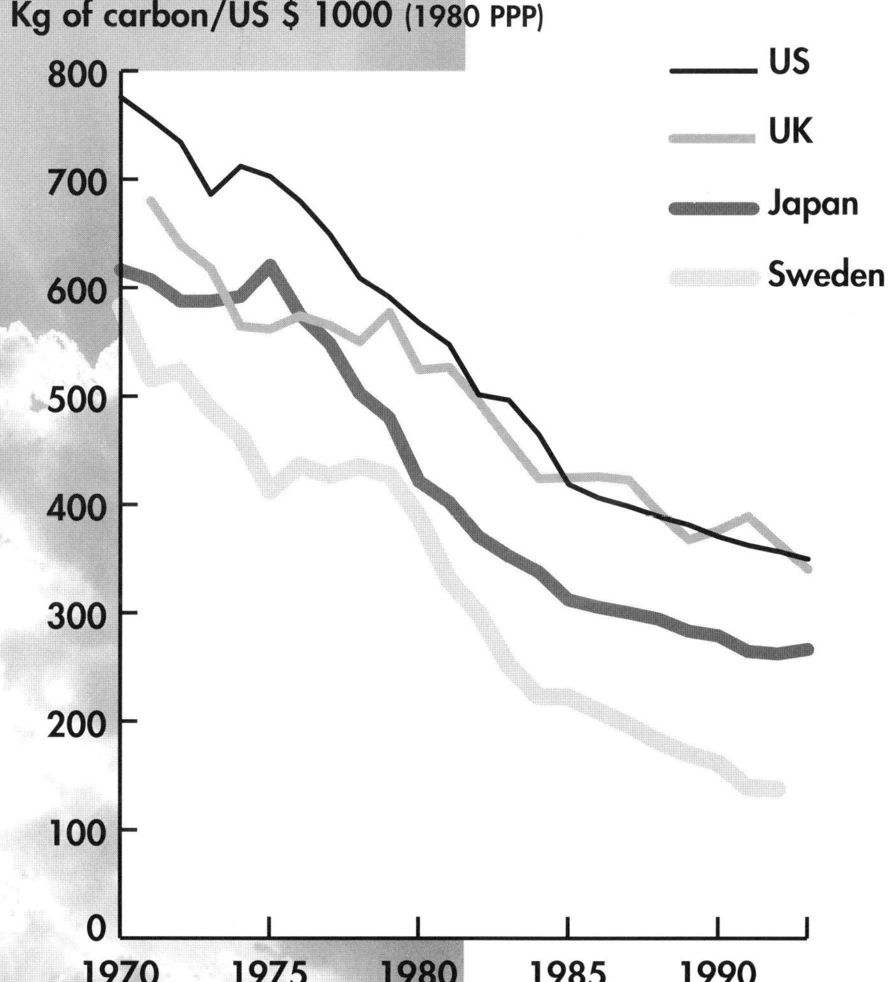

*** Holding structure constant at 1973 mix.**

Sources: National energy balances and IPPC co-efficients; Lawrence Berkeley National Laboratory.

Sweden, for example). A let-up in the rate of decline appeared in the early 1990s, possibly because of a decline in real energy prices. A recent IEA/OECD study on CO_2 and competitiveness shows that differences in carbon emissions per unit of output among IEA countries remain significant, with considerable potential for resuming the rapid decline in this indicator.

Indicators
of Energy Use
for Cars,
Trucks, Banks,
Schools
and Offices

TRAVEL, SERVICES

4

■ In contrast to the relatively optimistic news from the manufacturing and household sectors, other actors on the energy-saving stage – the travel, freight and services sectors – turned in rather poorer performances. This chapter covers all three but concentrates on the first, not only because of its weight in energy consumption and overall CO_2 emissions but also because existing information renders the travel sector most amenable to scrutiny with the indicators approach – and therefore most clearly illustrative of it.

■ Travel: Let's Take A Spin and Burn Some Carbon

■ Transport presents one of the biggest challenges for policy makers. It accounted for almost a third of total final energy consumption in IEA countries in 1995 and two-thirds of that came from personal travel. Over 80 per cent of it was road transport, which contributed an estimated 22.6 per cent of total CO_2 emissions in IEA countries in 1995. Attempts to address these concerns have focussed principally on technology to reduce fuel use per kilometre in new cars and ways to improve traffic flow. Some transport and environmental policies aim to get people to use cars less and switch to other modes such as bus or train; they appear to have had relatively little impact in most countries.

■ Using the full set of indicators for personal travel, this section looks at how they portray trends and policy impacts; it also offers a detailed look at the development of relevant indicators through the techniques of disaggregation and layering. Measuring changes in energy use, or differences in it among countries, requires an examination of each main component of the principal modes of transport: vehicle characteristics, usage, travel patterns and energy use itself. Technology, behaviour, energy policy, and policy instruments such as car and fuel taxation all provoke changes in how people travel and how much energy they use. Energy indicators bring out all these elements. A careful examination of non-technical parameters shows how such diverse factors as congestion, noise, fiscal policies and parking availability will affect travel; in turn, these factors will be affected by transport-related policies that do not address energy directly yet have a great impact on its use.

Structure

■ Structure here refers to physical and behavioural characteristics: how people travel and how far. During the oil crises of 1973 and 1979, policy makers exhorted drivers to garage their cars and use other modes of transport, or stay home. Except in these periods, however, few energy policies have aimed at the structure of transport itself. At the same time, structural changes have raised transport energy use significantly. Cars have developed more energy-intensive characteristics, for instance, even as policies aimed at fuel efficiency led to important improvements. Policy makers need indicators of all these complex changes to understand how energy use for travel has evolved.

Vehicle Characteristics

■ The stock of cars is the key element in travel or mobility. Car ownership per 1 000 people differed by a factor of nearly five between Japan and the United States in 1970. By 1993, the United States still led, but other nations had narrowed the gap.

■ Indicators of these characteristics clearly relate to fuel intensity or fuel economy. Data show an increase in average weight of new cars sold in several European countries and a rather striking drop in the weight of new US cars through the early 1980's. This came in response to both higher fuel prices and Corporate Average Fuel Economy (CAFE) standards. Car power and motor size grew in all the countries, nearly doubling by 1993 the ratio of power to weight in 1970 (though they actually *fell* in the 1970's and 1980's in the United States). In the United States the ratio of power to engine size, an indirect measure of how efficiently motors convert fuel to power, increased by half.

Vehicle Use

■ Distances driven in selected IEA countries show a wide spread and much ambiguity. The two European countries with the lowest rates of car ownership, Denmark and Finland, have the highest usage rates per car, and the two factors nearly compensate each other, measuring almost nothing. Measuring vehicle-kilometres per capita gives a far more accurate indicator of car use. Japan lies at the low extreme, with relatively few cars and, because of traffic congestion, the lowest distances driven per car; the United States occupies the other extreme on both counts. This indicator also captures the important effect of the load factor – the number of people per car. Load factors vary by the purpose of each trip and are generally lowest for commuting. The mean load factor has fallen in all countries as a result of the increasing numbers of cars, more commuting and smaller household size. Passenger travel in cars has increased less than vehicle-kilometres per car. Data also show that trips for work are now growing less rapidly than those for family and social purposes, leisure and vacation.

Fuel Choice

■ Vehicle energy use involves four basic fuels – gasoline, diesel, liquefied petroleum gas (LPG) and compressed natural gas (CNG) – as well as fuel additives, predominantly ethanol and methanol. Public policies can affect fuel choices. Several countries have created or supported significant differentials among the prices of gasoline, diesel and LPG. So measuring the impact of such policies on both fuel choice and fuel use becomes critical to policy making. This applies particularly to diesel, because drivers seem willing to pay more for a diesel car in order to benefit from its lower running costs. Market shares of diesel versus gasoline cars provide a ready indicator; it shows a dramatic rise of diesel cars in some countries. LPG cannot be neglected in a few countries where tax policy encourages its use.

Vehicle Energy Use and Intensity

■ Different kinds of vehicles move varying distances during the year, using fuel at varying rates relative to distance; so fuel choice does not translate into the actual shares of different fuels used. Total usage per 100 km by fuel and by car type can be measured by tabulating fuel use, distances driven and numbers of cars by fuel type. It measures the intensity of vehicle fuels or energy; it can also be used to measure vehicle kilometres per litre consumed or fuel economy. The figures are based on actual driving, and include diesel and LPG cars.

■ The fuel or energy intensities of car fleets vary significantly among countries, because of differences in vehicle characteristics (weight or power), traffic conditions and driver behaviour. Low mean temperatures clearly

Figure 8

Automobile On-Road Fuel Intensity

Weighted Average of Gasoline/Diesel Fuel Intensity

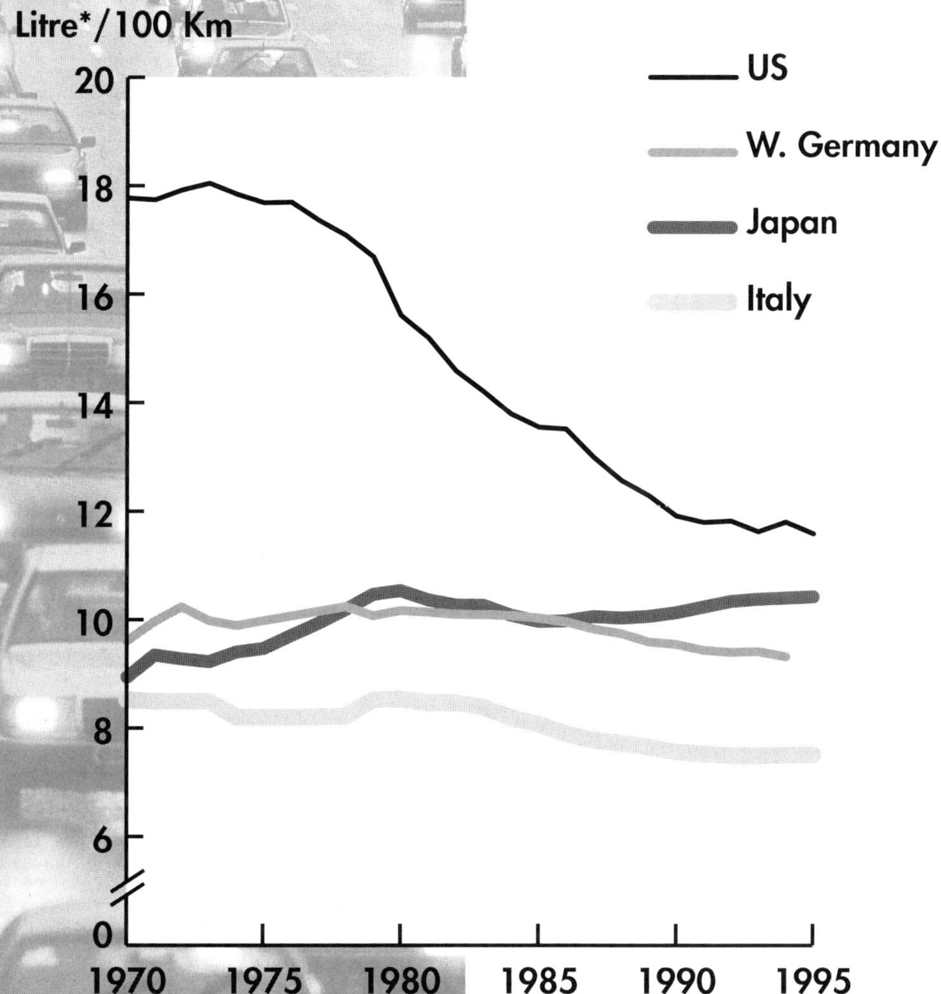

**Gasoline, Diesel and LPG included at energy content.*

Note: Data includes diesel, LPG for all countries; household light trucks for US, UK.

Sources: National transport statistics; Lawrence Berkeley National Laboratory; IEA, *Indicators of Energy Use and Efficiency*, (OECD, Paris) 1997.

increase fuel intensity in Canada and the Nordic countries in the winter. A 30 per cent drop in the fuel intensity of US and Canadian cars in the 1970's and 1980's contrasted sharply with limited declines in Europe and Japan (Figure 8). The values for the United States rapidly approached those for Europe through 1991, but the two regions have since diverged, with fuel intensity in North America again outstripping that in Europe.

■ The properties of new cars affect fuel intensity as more efficient cars enter the fleet and older, less efficient ones go to the junkyard. The sales-weighted new-car fuel intensity (or fuel economy) – that of each car weighted by its share in total sales – serves as a key indicator of fuel intensity at the margin, but only at the margin. New cars are actually driven farther and so factors such as distances driven, speed and traffic conditions resume their importance for actual fuel use. A short trip with a cold engine, for example, will require higher fuel use per kilometre than a long trip with a warm engine. These elements contribute to a gap of from 10 to 30 per cent between test fuel economy and actual fuel intensity.

■ Government policy approaches, whether voluntary or based the heavier hand of regulation, induced car manufacturers to try to reduce the fuel intensity of their new cars. Indicators did indeed register a fall in the period from 1970 to 1995 – but the lack of dramatic change in actual fuel economy suggests that the gap between actual and test values got forgotten in the policy discussion.

■ Dividing fuel economy demonstrated in tests of new cars by average car weight produces an indicator of improved aerodynamics, combustion and other efficiency-promoting features. Similar indicators using fuel intensity and horsepower also show significant gains. Motors provide more peak output (in kilowatts) per unit of engine displacement, another indicator of improved efficiency. And the ratio of horsepower to weight has increased, which means cars are easier to accelerate, an indicator of performance. Thus the indicator of fuel intensity, which has changed only slowly in most countries, understates real improvements in vehicle efficiency. Differences in fuel intensity reflect not simply differences in technology but also those in car features and traffic congestion. Although it is undeniably an important indicator of energy intensity, fuel use per kilometre is a mixed indicator that incorporates both technical factors and others such as usage patterns, lifestyles and incomes. To derive fair indicators of car energy use, therefore, the analyst must consider measures of use (in kilometres per year), together with load factors and one or more indicators of car characteristics, such as weight, that reflect consumer choices and affect energy use.

■ The fuel intensity of vehicle use divided by the load factor determines the energy intensity of car travel, or "modal intensity." The load factor has fallen by a third, so the intensity of car travel has increased by the same amount, all other things being equal. Indeed, when one includes the drop in load factor, one discovers a key link between energy and mobility – *more* energy required in the early 1990s than in 1973 to provide each passenger-kilometre of car travel in every IEA country but the United States and Canada.

Aggregate Indicators

■ Combining the main indicators outlined so far – per capita travel and modal intensities – yields per capita energy use for travel. Between 1973 and 1993, growth in car ownership furnished the main component of increased travel and its energy use in Europe and Japan. With per capita car ownership and use much higher than in other countries in 1970, the United States showed a growth rate considerably slower than in Europe and Japan. The fuel intensity of the US fleet declined more than 30 per cent from 1973 to 1993, while intensities in Europe and Japan fell by 13 per cent at most. Car use increased everywhere, above all in Japan. In the US, per capita fuel use for cars, though lower in 1993 than in 1973, remained considerably higher than in all other countries.

■ The trends can be decomposed to isolate per capita travel (activity), modal structure, and modal energy intensity. These three components are not necessarily independent of one another because reduced fuel intensity for a given mode can reduce its cost and thereby stimulate its use (this is the so-called "feedback or rebound effect"). Together, the three indicators explain changes in energy use for travel. Holding two of them constant and letting one vary shows how much it actually changed energy use. Increases in per capita travel boosted energy use from 1973 to 1992 by between 40 per cent (in the United States) and 65 per cent (in Japan). Shifts in modal structure raised energy use by 33 per cent in Japan, 3 per cent in the United States, and 4 per cent in Europe. Changes in modal energy intensity reduced US energy use by 18 per cent but slightly increased it in most European countries. The indicators reveal what aggregate figures cannot: that energy policy makers in IEA countries achieved their goal – to reduce energy intensity enough to offset growth in travel – only in the United States and Canada.

Differences Among Countries

■ The aggregated indicators suggest that any international comparison demands consideration of many elements, particularly if the aim is to transfer experience, policies, or technologies from country to country. Combining total travel by mode (cars, buses, trains, air) with modal intensities gives per capita energy use for travel. Cars dominate this indicator. Car ownership and travel, together with car design and fuel economy, are, in about equal measure, the factors that explain a three-to-one ratio between per capita fuel use in the United States and Europe in 1992. Combining motorisation and distance per car shows that the gap stems mostly from differences in per capita travel. A similar difference in fuel use per kilometre arises mostly from variance in car weight and size.

■ Do Europeans travel less in cars than Americans simply because their countries are smaller? National travel surveys for six European countries estimate the average car trip at between 13 km and 15 km; the US figure lies slightly below 15 km. Thus local travel predominates in car usage, regardless of country size or population density. Therefore, trip frequency, not length, determines total car use and hence total transport energy use. Higher frequency in the United States does have something to do with the country's large size, but the link takes the form of a proliferation of low-density settlements which favour cars over other modes. Detailed travel surveys show that many US car trips other than for work are shorter than some trips that Europeans or Japanese would make by foot, bicycle or local transit. More compact settlements do not so much shorten car trips as reduce their number in favour of other modes of transportation, and cheap fuel and parking as well.

"Driving" Factors in Driving: Why Energy Use in Travel Increases

■ What forces drive energy use for travel and what explains the marked differences in travel per capita among countries? Why have travel and related energy use increased so strongly? Policy makers repeatedly ask these questions because all the trends point to continued increases. Policies that lie outside the realm of energy policy or fuel taxation, especially fiscal and transport policies, have particular importance here. They can have a profound impact on total mobility or car use.

■ Rising income raises car ownership and increases travel. Energy use grows slightly faster than travel does, with the gradual shift to cars and away from trains and buses. In Europe and Japan a slow increase in the energy intensity of travel itself reinforces this trend, largely because load factors in cars fall faster than vehicle intensity. Shifts towards heavier, more powerful cars as average income climbs also play a role. Indicators point out that higher incomes also generate more car travel – i.e. greater use as well as more ownership. Drivers who benefit from company cars drive them farther than those who pay for their own fuel.

■ Income is not the sole determinant of what kind of car a family can afford. Taxation affects both ownership and vehicle characteristics. A comparison of prices and taxes for a representative model car shows enormous variation in Europe: Danish, Dutch and Norwegian buyers pay higher taxes and so buy fewer, smaller cars than people in other countries. US taxes are trivial by comparison. In some countries, however, company cars, a very lightly taxed employment perk, account for more than half of new cars

Does Market Saturation Apply to Cars?

The notion of saturation came up in Chapter Three with respect to household energy use. Does it have relevance for car travel? Data available to the mid 1990's suggest that saturation is approaching in the United States, Australia and western Germany. Car ownership per thousand people of driving age (18-70) is around 600 per 1 000 inhabitants in France and elsewhere in continental Europe, but reaches nearly 900 in the United States. About 75 per cent of French families have access to one or more cars, but about 90 per cent of US families do. A majority of US families have one car per licensed driver and at least two cars per family, while only 35 per cent of French families have two or more cars. These indicators suggest near-saturation in the United States but room for growth in France. Individual car ownership is a sensitive function of income, socio-demographic status and age. All measures show that the number of cars and share of the population with access to a car has grown in every country.

At the same time, car use per capita has not yet reached saturation. While it grows less rapidly than income in the most motorised countries (United States, Canada, Australia), some of that slowing clearly arises from substitution by air travel. Some say that time available for travel tends to be carefully budgeted. This may apply roughly to commuting, but not to all travel, including holidays. Moreover, a gradual increase in the speed of travel, affected both by increases in car use outside congested areas and times and by increased air travel, means that even without a significant increase in time spent travelling, travel keeps increasing.

sold. Swedish statistics reveal that company cars are heavier and more powerful than privately-owned ones. Implicit subsidies to commuters which ignore the mode of transport they use, as well as tax deductions for interest on mortgages which encourage people to buy larger houses on larger lots, also lead to more travel.

■ Along with incomes, prices have strong effects as well. With few exceptions, real fuel prices in the early 1990's were lower than in the 1970's or 1980s, in fact close to their values before the first oil shock. Combining fuel prices with fuel intensities reveals a fuel cost per kilometre no higher in 1992 than in 1973 in almost every country examined; this reflects both price movements and gains in fuel economy. Only Sweden and Canada had prices in the 1990's that were significantly higher than in 1973, in both cases because of serious shifts in taxation.

■ Income and price effects over the long term have indeed influenced people's travel energy choices, making the average car in Europe larger, more powerful, and heavier in 1992 than in 1973. Practically every car in Europe, Japan, and the United States produced more power and propelled more weight per unit of fuel consumed. Cars have become far more energy efficient in a technical sense, but much of this efficiency has gone to increase power and comfort – an important example of the feedback between energy efficiency and energy use itself. Sometimes stronger, sometimes weaker, these feedback effects act on many energy-using sectors.

■ Where do people go by car? Time-series data show rising travel and car use for family business, shopping, leisure and vacation, rather than for work. With rising incomes (and in the absence of effective policies which encourage the opposite), people abandon the denser urban environments that can support convenient transit service and walking. Other indicators point to the dependence of travel and car use on the socio-demographic composition of populations (income, family size, lifestyle, ages, number of workers in the household). The causes of changes in travel patterns go beyond incomes and prices to incorporate socio-demographic and geographic parameters as well as subtle features of tax policy. Not all can be reduced to simple indicators, but many can.

CO_2 Emissions from Travel

■ In every IEA country, policies aimed at restraining undesirable emissions focus on travel. Indicators of travel-related emissions thus become important complements to those that explain the underlying components of fuel use in travel. The first step in creating them is to tabulate emissions from each vehicle type, taking into account its energy use by fuel. After disaggregating diesel, gasoline and marine diesel, and taking into account the fuels used to provide electricity for railway trains, trams and trolley buses, the picture shows per capita emissions in the United States and Canada were at about the same rate in 1992 as in 1973, but are now rising. They remained three or four times as high as in Europe, where emissions rose steadily, closing part of the gap with North America. As figure 9 shows, emissions per unit of GDP fell sharply for a while in the US, while they rose slowly in Europe and Japan. This picture poses a clear challenge to policy makers, as it contrasts so sharply with the situation in manufacturing and households.

■ Changes among countries follow energy use, largely because emissions from diesel and gasoline, which dominate in all countries, differ little per unit of energy. It is

Figure 9

Carbon Intensity of Travel Relative to GDP

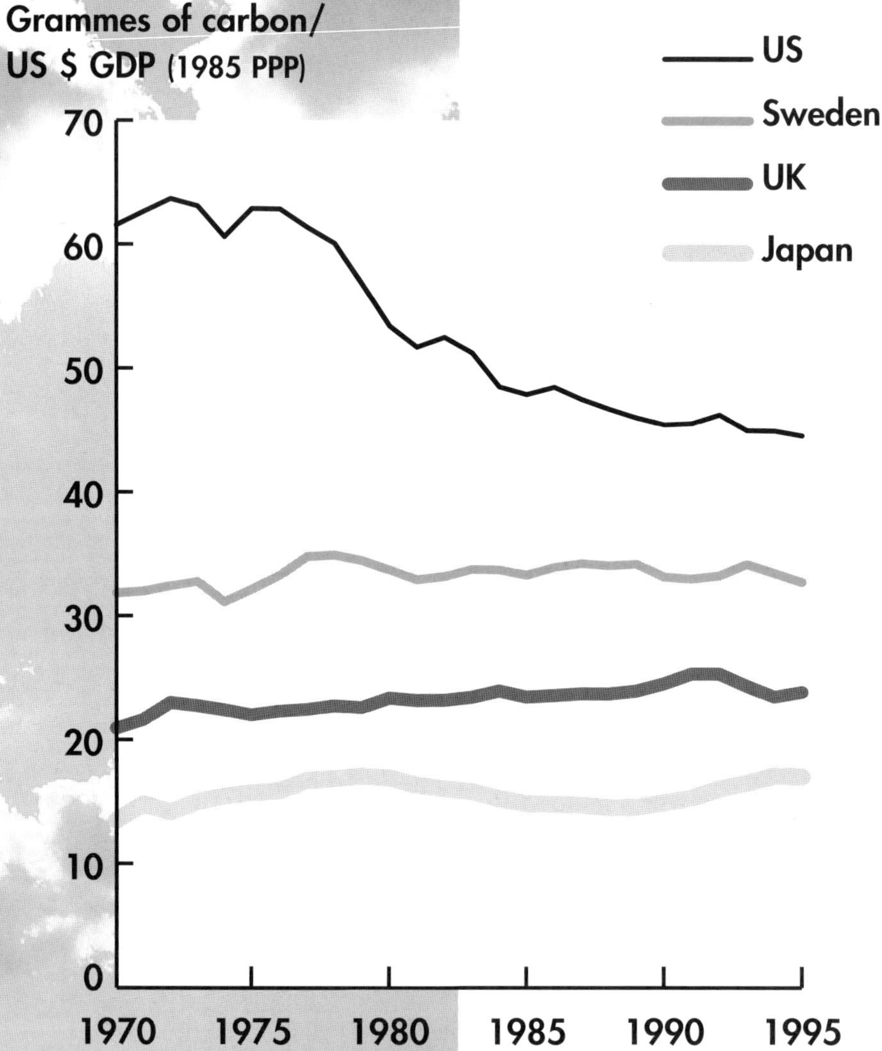

Sources: Lawrence Berkeley National Laboratory;
IEA, *Indicators of Energy Use and Efficiency*, (OECD, Paris) 1997.

hard to argue that the switch to diesel over the past two decades has reduced greenhouse gases or saved energy – so far. Technological advances in diesel engines may, of course, yet do so. (So far, the intrinsically higher efficiency of diesel engines has been offset by the lower price of diesel. Diesel cars are driven much farther each year than those using gasoline; so their total emissions are about the same.) A factor analysis of changes in the components of CO_2 emissions from travel mirrors the analysis of change in energy use. Adding a term to reflect changes in the fuel mix has only a slight effect, due to small amount of electricity now used in the sector. Changes in modal intensity in the United States and in modal structure in Japan, as well as overall increases in economic activity, are the leading components of changes in CO_2 emissions.

More Insights

■ Most policy issues related to energy and transport revolve around the efficiency of energy use, but measures of improved energy efficiency are difficult to derive. Indicators imply that cars now require less fuel to move a given mass a given distance. Cars have become more energy efficient, but their weight has increased and energy use per kilometre driven has not fallen very much. At the same time, per capita distances travelled have increased somewhat. Does this mean that car use is less energy efficient because it is more energy intensive? Furthermore, load factors have fallen more than enough to offset the small declines in energy intensities in Europe, resulting in an increase in the modal intensity of car travel. Does this important change represent "inefficient" behaviour? Similarly, switches away from non-motorised or collective modes of transportation have meant greater aggregate energy intensity for travel and higher energy use. Do these changes imply that energy use for travel has become less efficient? How much of the difference between consumption in the United States and in Europe or Japan is a function of differences in energy efficiency?

■ The indicators developed in the study suggest that the efficiency of cars in the narrowest sense now differs very little from country to country. It is hard to say how policies may have affected car efficiency, as the same effects appear everywhere, even in countries with no automotive industries or policies to alter car energy use. Differences in car weight and performance now cause the differences in fuel intensity among countries. And the indicators show that weight is rising. If fuel prices and new car taxes do determine properties such as weight, and hence fuel consumption, then local policies can affect fuel intensity directly by shaping the market for new cars; but they will not affect efficiency directly unless car makers respond by changing the efficiency of cars.

■ Has the travel sector saved energy? Only the United States and Canada exhibit significant deviations from historical growth patterns. In part, this is because they already had such high energy intensities in 1973. Another possible reason is that only in the those countries did government intervened directly to influence car design, through the CAFE fuel-economy standards. While cars' true efficiency increased significantly in all countries, increases in weight and power as well as worsening driving conditions in all but a few countries absorbed the impact of this improvement. Various agreements and targets set by government and industry committees, bolstered by strong research and development programmes in some countries, have brought about improvements in car technology, but these programmes only offset increases (or reduced declines) in fuel intensity that might otherwise have occurred.

■ Assigning causes to the measured effects is difficult and controversial: how much depends on price, how much on technology, how much on policy? Few analysts have unambiguous answers, but without indicators one can hardly even ask such questions. Travel presents a complicated and changing target. Energy policy makers must recognise this and be prepared to look to non-energy policies if they wish to be effective. Hence the many structural indicators introduced here take on increased importance in analysing key trends in travel energy use. The decomposition shows the danger of relying on simple figures – say, fuel use per capita or fuel use per vehicle – as a measure of energy efficiency or even intensity. The level and structure of travel have almost as much importance as the intensities of the different modes in determining overall energy use and explaining differences among countries. Even aggregate energy use per passenger-kilometre provides only an indirect measure of the efficiency of vehicles or intensities of modes. Equally, figures for aggregate fuel use for "road vehicles" do not say much about real developments in energy use for transport. The aggregates cannot measure past changes in the energy intensities of travel modes. They become useful indicators only with a full understanding of the components that go into them.

■ Freight: Roaring Down the Road

■ Typically, 10 to 12 per cent of final energy consumption goes into hauling freight in IEA countries, mainly in the form of oil products, with electricity providing a small share for electrified rail lines. Trucks, an increasingly important component of transport, dominate the energy balance of the subsector. Freight has encountered few energy-related policies, but likely will see them in the future because of pressures from transport authorities over congestion and noise, air pollution, and CO_2 emissions.

Structure

■ Freight has structural elements similar to those in the travel sector: the stock of vehicles, their characteristics and the distances they travel, the characteristics of freight (analogous to the purposes of trips), its quantity, and usage measured as freight haulage in tonne-kilometres. Also important are the choice of haulage mode (ship, air, rail, truck), fuel choice, fuel intensity and modal intensity. As with travel, all these elements combine to explain both changes in energy use over time and differences among countries.

■ Tonne-kilometres furnish the best measure of freight activity with GDP in goods-producing sectors the clear driving factor. Haulage- to-GDP ratios fell slightly in the ten countries studied between 1973 and 1992 as haulage rose (except in recessions), but GDP rose faster. Trucking gained an ever-increasing share of total haulage. Small or densely settled countries – Japan or the Benelux countries for example – generally register the highest shares of truck freight. The United States, Canada and Australia have the

Figure 10

Freight Activity and GDP

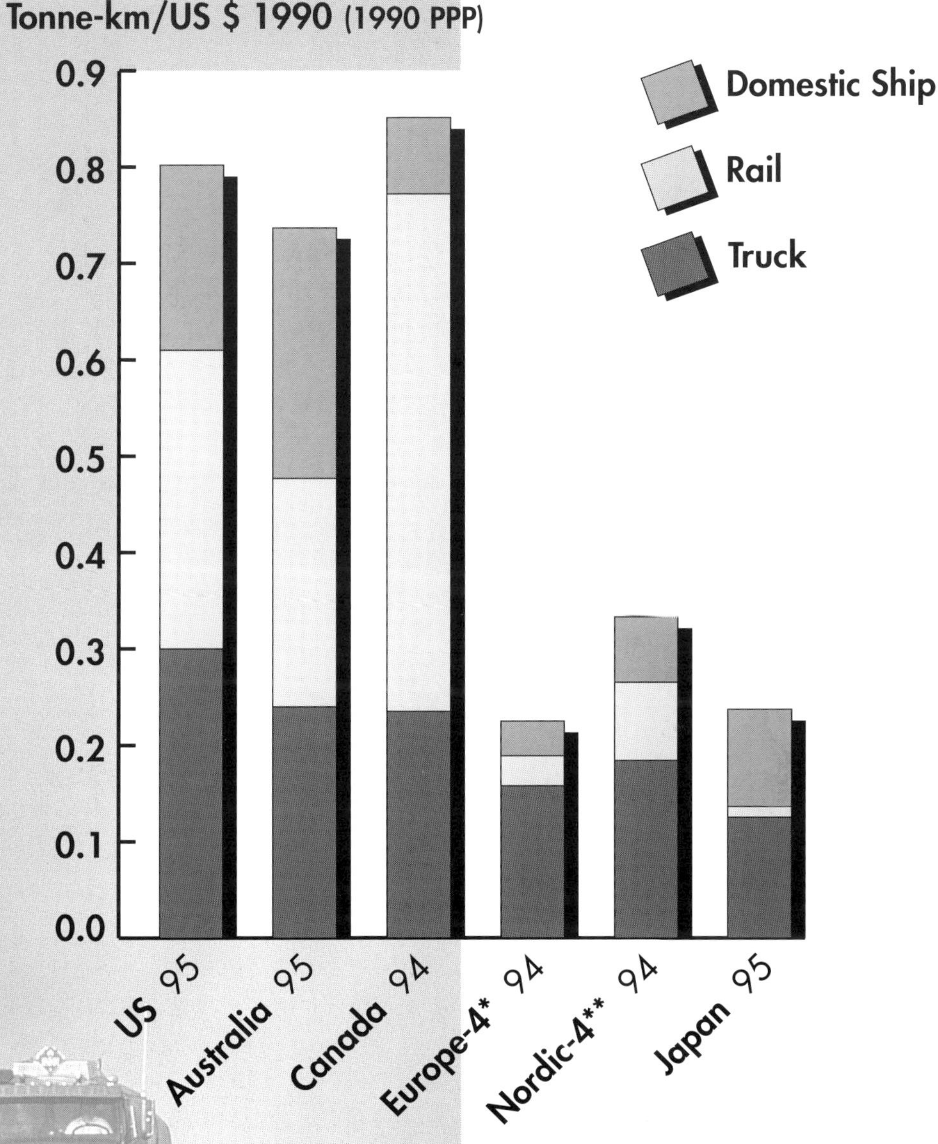

** Europe-4 = Italy, France, W. Germany, United Kingdom.*
***Nordic-4 = Norway, Sweden, Denmark, Finland.*

Sources: Lawrence Berkeley National Laboratory;
IEA, *Indicators of Energy Use and Efficiency*, (OECD, Paris) 1997.

highest ratio of haulage to GDP, followed by Finland, Sweden and Norway. These Nordic countries, somewhat larger and less populated than others in Europe, depend heavily on raw materials production. The United States not only produces many raw materials (ore, grain, coal) but also ships them much farther than domestic freight travels in other countries. Geography and raw materials distribution play large parts in the ratio of freight haulage to GDP.

■ Modal mix (figure 10) has a key impact on freight energy use because truck freight has a modal energy intensity ten times higher than shipping or rail. Since trucks offer greater flexibility, they assume an ever-increasing role in freight transport. Almost every country, but especially Japan and most of Western Europe, has seen a constant shift to trucking and away from rail. Economic output in the IEA countries has shifted from bulk materials towards smaller products congenial to truck transport. The growing use of "just-in-time" delivery in manufacturing also favours trucks.

Truck Energy Intensity

■ The mix of trucks, the loads they carry, the distances each kind of truck covers and the fuel intensities of individual classes of trucks all affect the energy intensity of trucking. A major shift towards smaller vehicles in the truck fleets of every country has tended to increase modal intensity. Underneath these comparisons, however, lie many fluctuations in energy intensity and important changes in the nature of trucks and trucking. Changes in usage offset small but measurable improvements in the energy efficiencies of vehicles themselves. For most countries, the aggregate energy intensity of trucking hardly changed (although it showed inter-year fluctuations) over the period from 1970 through 1994.

■ The shares of trucks by weight class, the distances they drive and their usage all help to explain changes in the energy intensity of this important transport mode, as well as differences in intensities among countries. Vehicle intensities within weight classes relate closely to vehicle efficiencies, but trucking mix, usage and traffic conditions play greater roles than energy efficiency. They also explain differences in intensities among countries. It is a conundrum for energy policy makers that non-technical factors have a dominant influence on the modal and energy intensity of trucking. Few of them lie within the realm of energy policy; transport policies and other non-energy policies – or no policies – touch them.

■ Aggregate energy intensity of freight transport (energy use per tonne-kilometre) increased from 1970 to the mid 1990s in the majority of countries, by amounts ranging from 4.5 per cent in the United States to 28 per cent in Japan. Modal shifts contributed strongly to the increase in Japan, more than offsetting a decline in energy intensities from other factors. For each country, the structure effect is more important than the intensity effect: the increase in energy use produced by the shift to trucks exceeded on average any reduction in energy use resulting from changes in modal energy intensities. Trucking has absorbed most of the increase in overall freight activity.

■ The energy intensities for trucking vary among countries for many reasons that have little to do with fuel use technology and much to do with usage. Modal mix, capacity utilisation, and the overall volume of freight relative to GDP or population are the three most important indicators explaining differences in energy use per unit of GDP or per capita for freight. Vehicle energy efficiency plays only a minor role, largely because vehicles are produced in an international market.

Figure 11

Carbon Intensity of Freight Travel

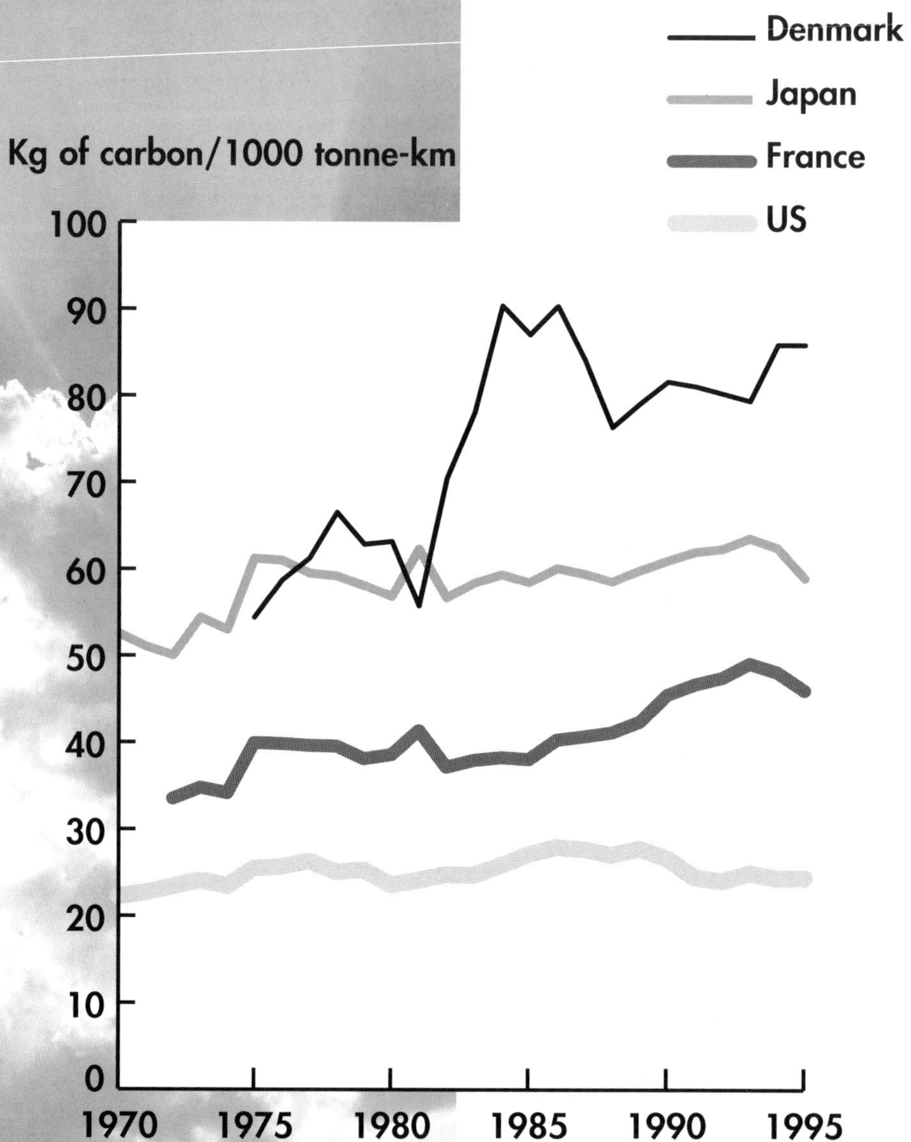

Source: Lawrence Berkeley National Laboratory.

Driving Forces

■ Three factors have driven energy use for freight: (1) the nature of trucking – the stock of trucks, their usage and fuel intensities independent of the technical properties of individual models of trucks; (2) regulatory conditions that vary by country and affect modal mix and modal energy intensities; and (3) the mix of trucks in the fleet, along with traffic conditions. The most important changes in energy use for freight relative to GDP or population depend on economic or managerial rather than technological influences. Indicators of energy use and efficiency must carefully document these non-technical factors but not neglect technology. Most freight-carrying technologies became more energy efficient even as other effects on overall energy use overwhelmed them. Without the new technologies, energy use for freight might today be as much as 15 per cent higher.

CO_2 Emissions from Freight

■ Because energy use increased, total CO_2 emissions from freight grew significantly in all countries examined (figure 11). Both modal mix and energy intensity contributed to the increase. Modal shifts pushed up emissions per tonne-kilometre, though the effect was offset in a few countries by falling energy intensities. Differences in fuel mix among countries are small and have little impact. So far, the coupling between GDP and freight energy use shown by the indicators has meant increased CO_2 emissions. Yet here again the indicators reveal many targets for change, whether in technology or in behaviour, to cut energy consumption and CO_2 emissions.

■ Services

■ The service sector includes activity in offices, hospitals, schools, concert halls, shops, warehouses, assembly buildings, and many others. Ill-defined in many energy statistics, it is often credited with whatever cannot be attributed to households, manufacturing or transport. Historical studies complemented by more recent and reliable information from government surveys, engineering studies, and studies by utilities of their billing market experience provide a partial view. For most countries, a lack of data limits the scope for disaggregated analysis of the underlying components of energy use in the service sector. In practice, the only uses that permit this type of analysis are space heating, cooling and lighting.

■ Measures of per capita GDP reflect activity in the services sector fairly well. But a country's overall building stock (floor area expressed in square metres or square feet) propels energy use more directly; it has shown steady per capita growth. The share of area by building type (retail, office, health), which varies by country, is important because each type has its own intrinsic energy consumption patterns which depend on the kinds of equipment installed. Usage can be estimated from the number of hours buildings are open or the number of employees per square metre, or in specialised measures such as meals served (at restaurants) or patient-days of care, all of which themselves vary enormously by building type or over time.

Energy Intensity

■ Primary energy intensity, although it shows fluctuations, has not changed much in most countries. Space heating has remained the dominant energy end-use in the service sector; electricity grew more rapidly here than elsewhere over the past three decades. Where data are available, they show that the use of oil and gas for heating declined relative to the area heated. The ratio of fuel consumed to floor area gives an approximate indicator of space heating efficiency, and it has shown a strong decline in most countries. This suggest real heat savings, and surveys confirm this. Electricity intensity, by contrast, is rising, with the growing presence of computers and air conditioning in addition to motors, lights and other uses. Indirect evidence suggests that, while the efficiency of each individual electricity use has improved considerably, the number of uses has grown more rapidly, leading to greater electricity intensity even without taking electric heating into account.

Aggregate Indicators

■ Aggregate indicators of energy intensity do not sufficiently separate the effects of changes in the sector's structure, efficiency, and end-use service or usage. A good understanding of energy use in the sector requires segmenting it into building types and end-uses or, alternatively, into economic activities and then into building types and end uses. The sector covers very different building types with very different end-use service requirements; new technologies and end-use services penetrate different building types at different rates. Therefore, only such segmentation makes it possible to talk about the effects on energy use of structural trends, trends in end-use service demand, and efficiency improvements. Lacking these valuable details, policy makers will miss seeing changes in the patterns of energy use, particularly those related to specific technologies, efficiency of energy use, or other factors that energy policies might influence.

Figure 12

Service Sector Space Heat Intensity

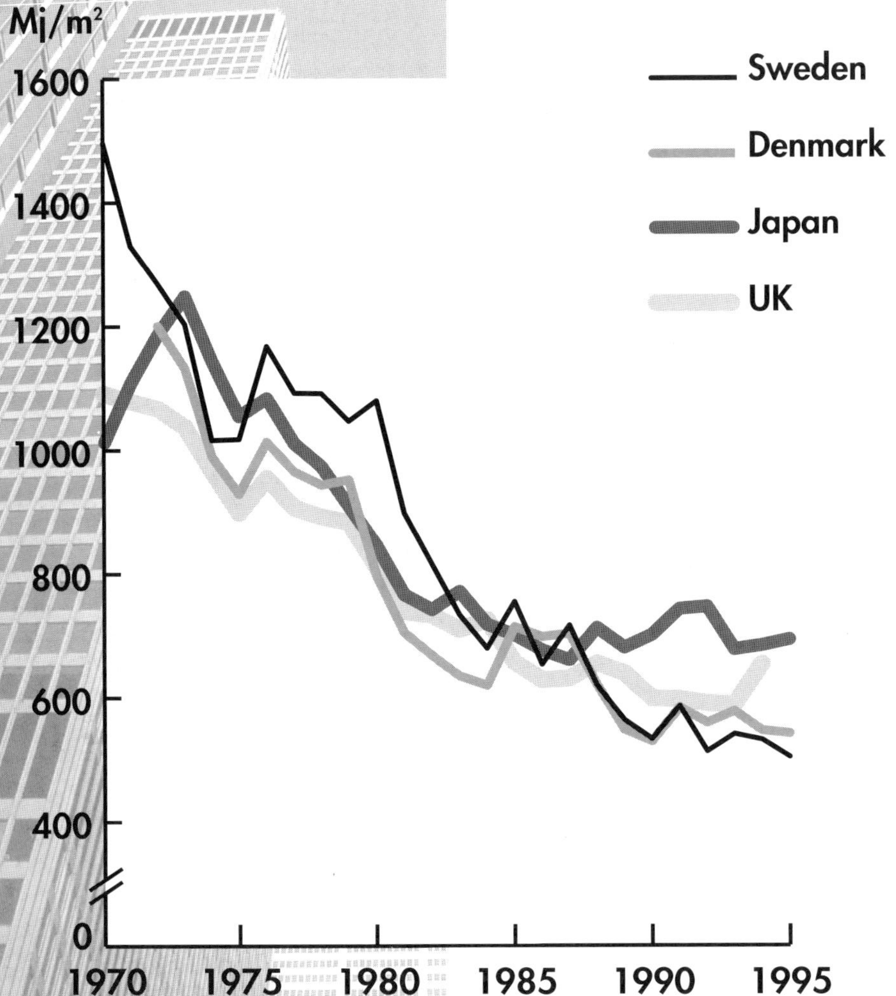

Source: IEA, *Indicators of Energy Use and Efficiency*, (OECD, Paris) 1997.

■ Nevertheless, the aggregates can help. By combining the fuel and electricity intensities, one can look at delivered energy use, normalised to area (figure 12) or output. These ratios fell in every country, and primary energy fell in some. When either primary or delivered energy is related to population rather than output, however, the resulting indicators show little change or even increase, reflecting the steep rise in services GDP in almost every country. All of these aggregates summarise the behaviour of service sector energy use but they give little insight into the efficiencies or intensities of any particular end-use.

■ Examining service sector energy uses across countries entails problems, whether in the aggregate, by building or by end-use. To get a better understanding of actual differences in efficiency, one has to look at the shares of different equipment and services, and at fuel shares in final energy use. Certain services may require substantially more energy than others. One cannot draw conclusions about the energy efficiency of the service sectors of countries merely by looking at aggregate energy use trends.

■ Because of such uncertainties, interpretations of changes in energy use scaled to area or output are potentially misleading. The rise in primary energy use per dollar of output means that the energy resource cost of that output has also risen. This could reflect either more machines, computers, lights and other devices used in businesses or less efficiency in particular cases. Significantly, however, the ratio of primary energy use to services GDP has fallen in most countries. With the enormous increase in the number of electricity-based end uses in buildings, this indicates improvement in the efficiency of each end use.

■ On balance, the indicators for services are inconclusive: heating intensities have probably fallen, but the changes in electricity uses defy interpretation, because no way exists to measure services accurately. Even stripping away electricity used for space heating leaves an unruly aggregate which will not aid in forming policy measures that could change the efficiency of energy use.

CO_2 Emissions from Services

■ CO_2 emissions from the service sector resist disaggregation by end-use except for certain countries and years. While emissions generally fell relative to services-sector GDP, another indicator – total emissions per capita – did not. In many countries the emissions from electricity used have risen and now dominate the total. As in other sectors, economic activity per capita generates the most important differences in per capita emissions among countries as well as substantial differences in emissions per unit of output. The high share of electricity means both that emissions may be disproportionately high, and that they can be reduced either by using primary electricity sources that are low in CO_2 or by moving to very efficient gas-fired generation of electricity.

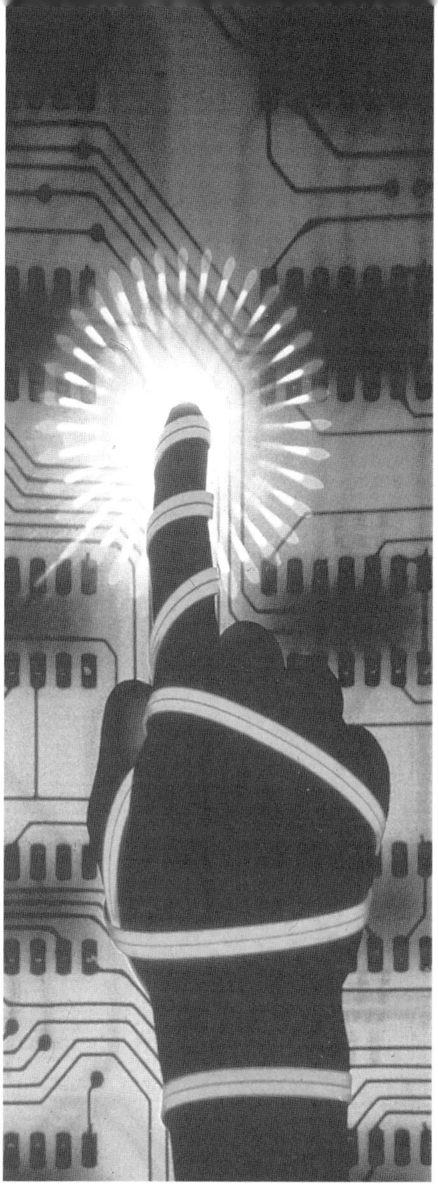

Causal Indicators: The Drivers of Energy Use and Energy Efficiency

DRIVING FORCES

5

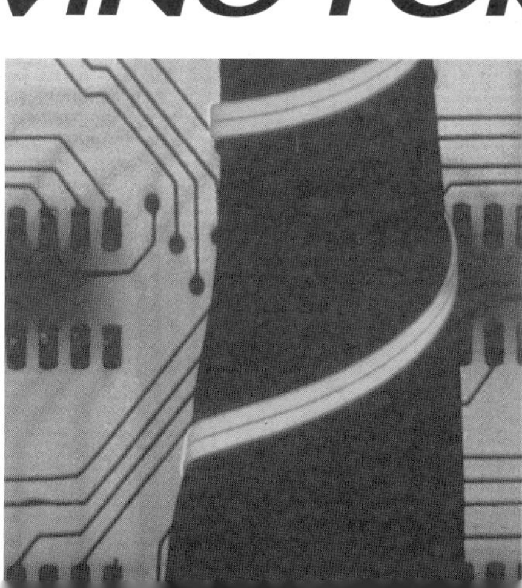

■ The preceding two chapters give a flavour of the information and insights that the indicators system can lend to monitoring energy use, understanding its origins and developing effective policy. This chapter stands back a bit and pulls together some of the principal conclusions, especially those concerning the reasons why people and institutions use energy, and use it in changing amounts. It will also consider the relative importance of those reasons.

■ Energy end-use indicators offer a technique for measuring changes over time. They represent one way of decomposing changes in aggregate and sectoral energy uses into components related to sector structures and end-use energy intensities. Built up from disaggregated data, the indicators provide a link between bottom-up and top-down approaches in examining the evolution of energy end-use. Indicators further establish quantitative links between energy end-use patterns and CO_2 emissions. The method improves understanding of recent emissions trends and differences in emissions profiles among countries. It offers an informative path to analysing efforts to mitigate future emissions.

The Past As Prologue? Not Necessarily

■ The exhaustive quantitative study in the IEA's *Indicators of Energy Use and Efficiency* reveals sometimes surprising patterns of energy savings in the IEA economies over the two decades up to the early 1990's. Some sectors performed much better than others; efficiency improvements came rapidly and naturally. Manufacturing and households emerged as the two "stars", the first propelled by energy prices, technical change and the discipline of competition; the second moved also by prices, but with a boost from efficiency standards and conservation programmes. Both showed considerable long-term restraint on energy use and CO_2 emissions, a characteristic which was accentuated by the oil price shocks of the 1970's, but which persisted even when real energy prices fell back to their pre-crisis levels.

■ Do these developments reflect lasting behavioural changes by energy decision-makers at the grass roots? Perhaps yes; perhaps no. Developing and watching good indicators will provide a more conclusive answer. Developments to date may simply reflect long structural lags between stimuli – price shocks – and responses. Much energy saving has come from technical change in the capital goods and consumer durables that use energy. Such equipment is long-lived and follows long investment cycles. In manufacturing, restraint began well before the 1970's and the first oil crisis (although it has shown a worrisome if still inconclusive tendency to slow in the 1990's) . On balance, behavioural predilections towards saving energy and using it more efficiently probably are ingrained more deeply in the manufacturing sector than it has been given credit for. In households, behavioural change clearly has occurred. Economic forces and public programmes have combined to make the average rich-country householder more energy-conscious and a better-educated customer today than he or she was 25 or 30 years ago.

■ Travel and freight present a different case altogether. Although some countries have shown small energy savings in the freight sector, freight activity has constantly climbed with GDP. Outside North America, no country showed significant savings or even restraint in energy use for personal ground transportation because changes in automobile fuel intensity proceed at a snail's pace. The two subsectors of transportation (trucking and automobiles) need policy focus. Neither has changed its behaviour at the micro-level of energy decision making.

■ Yet freight, especially trucking, as a commercial activity run by practical business people, should respond to the same complex of economic incentives as has manufacturing. As to personal travel, almost every householder also now owns one or more automobiles, but the virtuous homeowner who fastidiously turns out every light to save energy often guzzles gas like a drunken sailor when he gets behind the wheel.

■ A potential for change exists. The United States demonstrated it with its effective CAFE standards; European drivers show it in their demand for diesel cars to reduce running costs, even if this does not do much at the moment for the emissions problem. In the face of such potential, government transport policies seem inattentive or even perverse. Much of the problem stems from lack of co-ordination between energy and transport policies, made by ministries innocently at odds with one another because they do not communicate sufficiently.

■ Services show a mixed performance. Great fuel savings continue, as do improvements in electricity-use efficiency. Yet the indicators also reveal a rigid relationship between the sector's electricity use and its output (GDP); an exact coupling depends on climate, the price of electricity and the penetration of electric heating. Many studies rate the potential for electricity savings here as high and relatively cheap, with technology the primary vehicle of progress. The sector is a good candidate for carefully structured efficiency policies focused on electricity use. Key players include suppliers of building equipment, architects and engineering firms, associations of building owners and managers and, in many cases, of tenants as well. Once again, we see the same business and consumer interests that have done so well in saving energy in manufacturing and in households. Governments, which typically control a significant portion of the building stock, have no excuse for not being in the vanguard of energy savers in this sector. Indeed, their presence in it opens a large opportunity to show its further energy-saving possibilities, creating a "demonstration effect" that could save public money, reduce emissions and spread rapidly throughout the private segment of the sector.

■ Summing up the sectoral effects, *energy use and emissions are rising*. The most important factor behind rising emissions is economic growth, but a slowdown in the rate of improvement of energy efficiency is a key factor as well. Current trends do not point to a strengthening of the elements that restrained fossil fuel use and CO_2 emissions over the past two decades. Nor, however, do they necessarily point the other way. With proper policies, favourable basic trends which still exist can accelerate and unfavourable ones be blunted. If the indicators system described in this book is to earn its keep, it must prove itself better than other methods in identifying both the opportunities and the obstacles.

Key Forces at Work

■ In a very long-term perspective, the most important forces affecting energy demand include, roughly in order of importance:

- incomes or output;
- energy prices and other costs associated with using energy;
- technological change;
- social and demographic forces (workforce participation, family size and age distribution, life expectancy);
- and geography.

■ The exact role of each depends on the sector, the time period, and the level of development; saturation is an important potential factor. Most IEA countries have achieved widespread car and home ownership, as well as basic comforts in commercial and public buildings. Future energy services may focus more and more on low-energy, high-value electronics, but equally on additional transportation. Each country has to take stock of where its economy lies along these evolutionary paths in order to anticipate how much more growth may occur in these services, and hence what level of CO_2 emissions to anticipate.

■ Rising incomes clearly lead to rising energy use, but long-term elasticity (the per centage increase in energy use in response to a 1 per cent rise in income) becomes less than 1 after a society has built its basic energy-consuming infrastructure. The long-term carbon intensity of most countries' fuel mix has fallen. Both the evolution of the economy and changes in fuel mix tend to yield a progressively lower ratio of CO_2 emissions to GDP. Since the mid 1980s in many countries, however, growth in GDP has exceeded this "decarbonisation" rate, so that overall emissions rose. Policies to reduce energy use or CO_2 emissions should recognise which forces (particularly those outside the energy or environment policy-making sphere) lead to more or less decarbonisation, and indeed which forces (primarily in transport) may actually promote increases in the carbon-intensity of the economy. Good CO_2 policy harnesses the power of these non-energy related policies and seeks to circumvent those policies that may be raising emissions. Recent trends in transport policies, for example, may slow the rise in goods and personal mobility for reasons not related to CO_2 and thus indirectly contribute to emissions restraint.

Figure 13

Changes in Key Indicators of Energy Use and Efficiency*

1973-1994

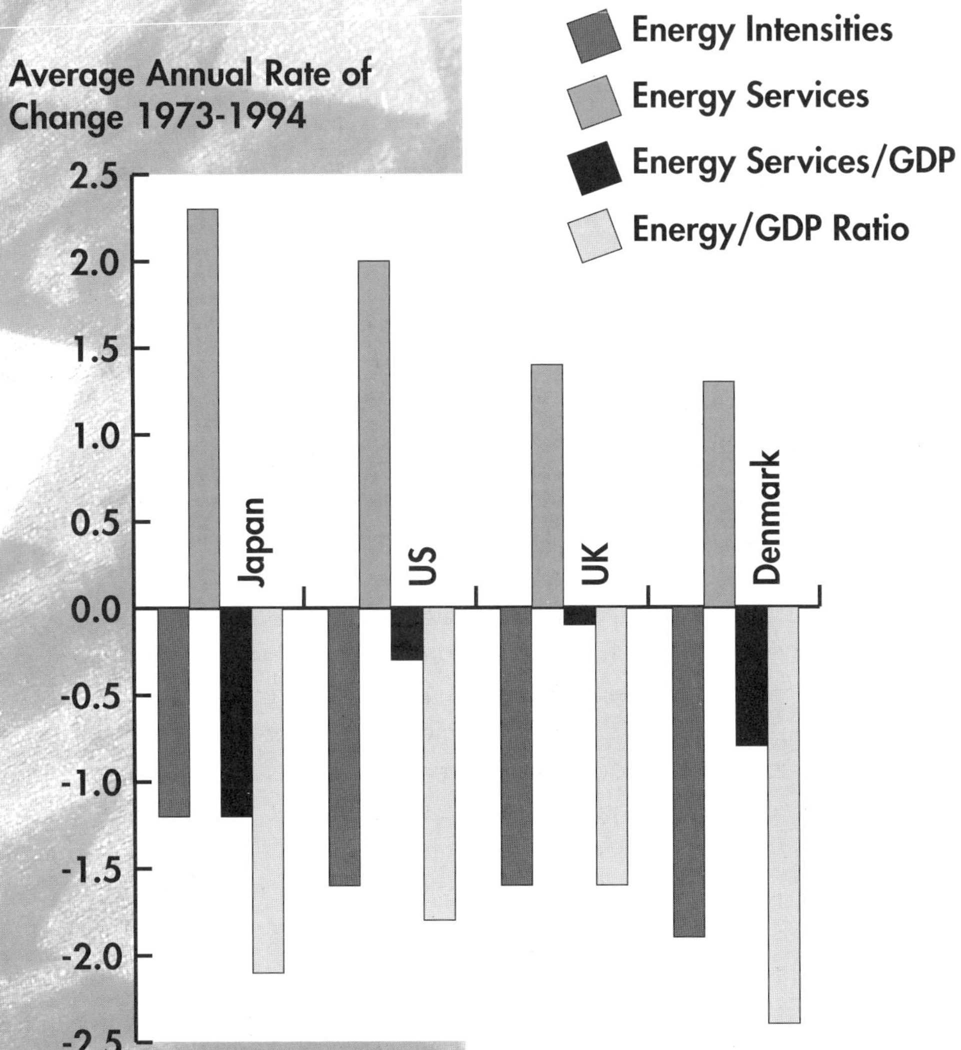

*** Based on 1990 economic structure.***

Source: IEA, *Indicators of Energy Use and Efficiency*, (OECD, Paris) 1997.

■ Figure 1 showed how overall changes in energy intensities reduced consumption and led to significant savings. Other factors worked to change energy use as well, especially structural changes in the mix of sectoral activities and increases in economic activity itself. Both factors increased energy use, as the bar labelled "energy services" in Figure 13 shows, but generally not as fast as growth in GDP. Consequently, the ratio of energy use to GDP often fell significantly more than did energy efficiency or intensity, as Figure 13 demonstrates. This comparison illustrates why the ratio of energy use to GDP, taken alone, is a poor measure of a country's energy efficiency and inaccurately measures changes in it. Because policies generally aim to reduce energy intensities, policy makers must be able to observe changes in intensities separately from changes in other factors. "Energy services" provides another valuable indicator of how the outputs for which energy is used change.

■ Forces driving improvements in energy efficiency include energy prices, long-term technological changes which are relatively independent of energy prices and energy efficiency programs. The first two forces drove most of the changes that occurred between 1973 and 1993. The many good examples of energy-efficiency programs or policies that had a profound impact on energy use (thermal building codes for new homes, the CAFE standards on new cars and light trucks in the United States, for example), acted on only a relatively small part of the total matrix of energy use in IEA countries. Many indicators permit comparisons among average technologies, average new technologies, and the least energy-intensive new technologies. These comparisons suggest significant further potential for energy saving.

■ Energy prices act on the rates of decline in energy intensity and on the adoption of new efficiency improvements; they also help to determine what "cost effective" savings are available. The replacement of old stock by newer, less energy-intensive homes, vehicles, appliances and capital goods automatically reduces energy use. But the rate of investment and the amount of reduction depend on energy prices. "No-cost" energy savings do occur as stock is replaced and new technologies invented, but overall savings in individual end-uses clearly come more rapidly with higher energy prices, even if the lags are significant. Thus whatever the "potential" for energy saving, prices sooner or later play a key role in what actually happens.

■ Figure 14 shows the relationship between the price and consumption of car fuel. Consumption in low-price countries is much greater than in high-price countries, mainly because of greater driving distances but also because fuel used per kilometre is 20 to 40 per cent higher in countries where gas is cheap than in those where it is dear. Similarly, Figure 15 shows that the use of fuel or electricity for heating and cooking, relative to GDP, tends to be high in countries with low prices. This is particularly true of electricity, which is used only sparingly in countries where it is expensive but is used lavishly in low-price countries.

■ The rate at which increased energy efficiency increases demand for energy services appears relatively small (a 1 per cent improvement in efficiency leads to less than a 0.2 per cent increase in energy service demand). Thus one can expect energy savings from programmes and stock turnover without a major bounce back in energy use. This does not apply in low income countries or among low-income consumers in IEA countries. In both cases a significant latent demand for energy services remains unmet because of constraints on income or supply, and because greater efficiency tends to boost the use of those services.

Figure 14

Automobile Fuel Use versus Average Fuel Price*

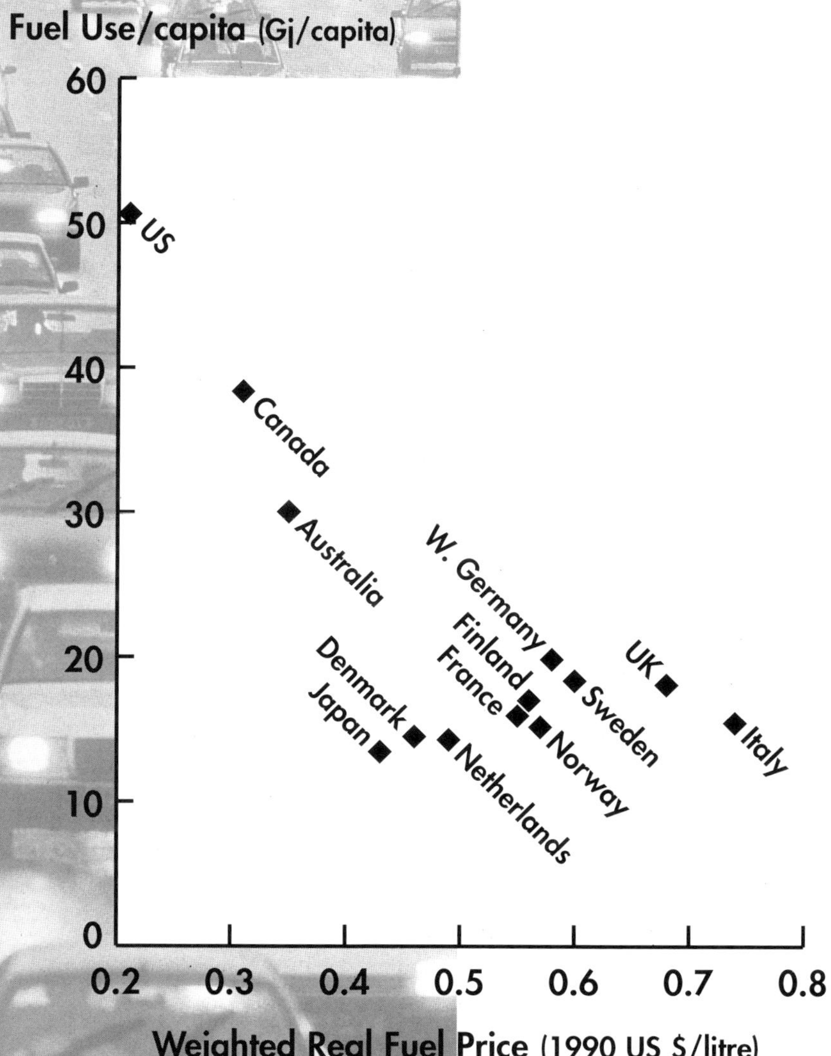

** Includes diesel, LPG used by cars and household light trucks.*

Source: IEA, *Indicators of Energy Use and Efficiency*, (OECD, Paris) 1997.

Figure 15

Household Energy Use Relative to Income

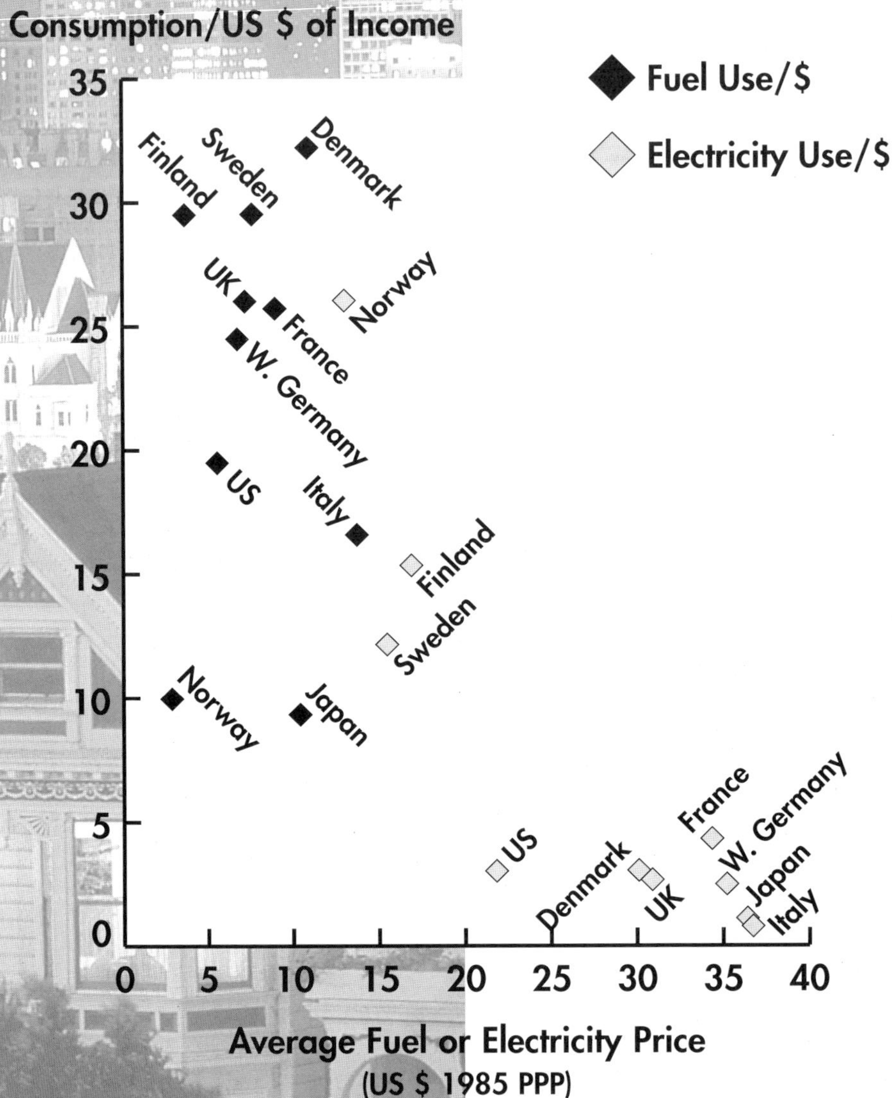

Source: IEA, *Indicators of Energy Use and Efficiency*, (OECD, Paris) 1997.

■ Cross-sectional differences among countries' per capita CO_2 emissions decompose into differences in income or per capita GDP, differences in the structure of energy use (house size, distances driven or steel produced relative to GDP), intensities of each energy use, final fuel mix and utility fuel mix. Aside from per capita GDP, which is the biggest component everywhere, the relative importance of each of these factors varies from country to country. For example, the indicators show that the ratio of freight hauled within a country to GDP is the most important determinant of fuel used for freight. The mix of modes (truck, rail, inland ship, and air) comes next. Fuel mix is relatively unimportant. By contrast, in the household sector, fuel and utility mix have as much importance as structural factors.

■ Figure 16 shows how the carbon intensities of several economies have changed. These changes reflect shifts in the carbon intensity of each activity (home heating, passenger travel by mode and so on), weighting each by its relative importance to emissions in a single year. The indicator represents the results of energy saving and fuel switching in the past, but it will be invaluable for monitoring or projecting future impacts on emissions of policy-induced and autonomous changes in carbon intensities. A major policy debate today concerns the rate at which this indicator, which has declined for many decades, will continue to decline even in the absence of policy action. A closely connected question is: how much further can policy actions push the individual components downward?

■ A distinct shift has occurred in the use of energy – from production (industry, agriculture, freight haulers and large enterprises in big buildings) to individual consumers and smaller businesses. Defining the "consumer" sectors as households, personal travel, and roughly half of the area in buildings where consumers go to buy goods or services

Figure 16

Impact of Changes in Carbon Intensity on Overall Emissions

1970-1994

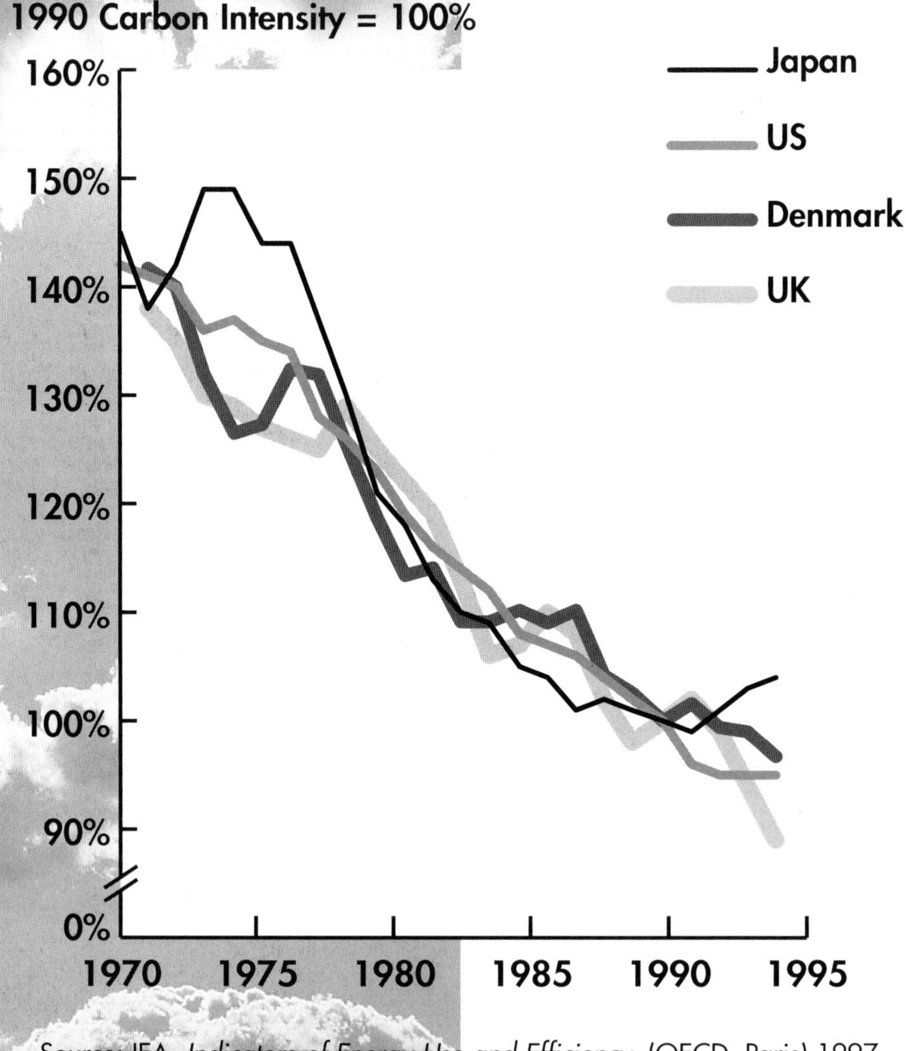

Source: IEA, *Indicators of Energy Use and Efficiency*, (OECD, Paris) 1997.

(hotels, restaurants, schools, hospitals), the share of energy consumption "for pleasure" is rising while the share for production falls. Because consumers, unlike businesses, do not have competition incentives to save energy, the calculus of energy saving changes, quite possibly weakening the forces driving savings. This possibility may justify seeking good administrative and regulatory approaches to efficiency, to complement energy pricing; such approaches include efficiency standards on buildings and equipment, information and advice.

■ The share of electricity as a fraction of final consumption in every country has risen steadily but appears to be approaching saturation. Countries with low prices (Norway, Sweden) have high shares due to space and water heating and a shift in manufacturing towards more electricity-intensive processes and products. Other countries such as the United States have lower shares because fossil-fuel heating remains much more prevalent.

The Indicators' Role in the Policy Process

POLICY MAKING

6

■ In a nutshell, indicators can show policy makers what they need to know to make good climate policy and follow up on what goes right or wrong. They fulfill a vital information function.

The Policy Process

■ *Problem identification* is the first conceptual step in any policy formulation process. In terms of effective policy, perceptions that climate change exists and that carbon emissions from human activity largely cause it really amount to an empty box. The box contains no real policies and insufficient information to make them. A system of indicators can reveal the present structures of emissions and their key underlying trends. More than that, it can identify the locus, the nature and the force of human decisions about energy use which create the structures and the trends.

■ *Goal setting* comes next. Politically acceptable goals can be a powerful organising force for moving societies towards co-operative ends. Emissions goals based on bad information will probably not be met and could generate popular opposition because those who make energy-use decisions at the grass roots can recognise the unrealistic. On the other hand, goals established with good information stand better chances of realisation and are more likely to gain political acceptance, whatever policy measures may be taken to reach them.

■ Indicators can facilitate realistic goal setting. Perhaps here above all the "bottom-up" techniques of an indicators system become highly relevant. Not only do they add precision to aggregate goals but also, and more importantly, they permit clear, unambiguous identification of the levels of activity at which goals can most appropriately and effectively be set. In any sector or for any of the dozens of key energy-using activity, some appropriate level between the aggregate and the individual human decision will exist for establishing an emissions objective. Bottom-up procedures help greatly in finding it.

■ *Policy design and implementation* now come into play. What measure can best meet the goals within the limits of political acceptability? What measure can be tailored specifically to the objectives, at the levels at which the objective are set? A decision to use fiscal policy - taxes, tax incentives or subsidies - may be appropriate at the aggregate or macroeconomic level if the target is also an aggregate one. Other kinds of policy action, whether regulatory or voluntary, based on research and development or designed to improve public information, sometimes have more flexibility and may be effectively targeted at various levels.

■ Having already helped to identify problems and their locus and to set goals at the most appropriate levels, indicators will have said much about the most appropriate kinds of policy measures to adopt and where to apply them. The indicators exercise also suggests that effective energy-use and restraint programmes should combine price signals, research and development and efficiency measures. At this stage, the focus turns even more towards the activities which policy is to influence. Because the indicators begin with activities, they have still another role at this point.

■ This book's survey of what the indicators say about five key sectors has uncovered several differences among them. The good energy-saving performance in households and manufacturing over recent decades suggests that policies should aim at accelerating basic trends already in evidence in those sectors; the same applies to services, for somewhat different reasons. In transport, personal travel and freight, the indicators point to several facets. Mobility has as much importance as fuel intensity in both. Fundamentally, mobility belongs to transport policy. Energy policy makers must marry their efforts with the broadening trend among transport authorities to address growing problems of noise, pollution, congestion and road capacity. They must co-ordinate their efforts to restrain energy use and CO_2 emissions with transport sector reform policies. This could involve bringing regional and municipal authorities more deeply into energy policy.

One Demographic Indicator

Consider just one example. Practically all of the IEA economies currently face the prospect of population ageing and all of the major economic and social issues that ageing raises. A graying population also poses the issue of future automobile use and fuel consumption. Data from the US Department of Transportation indicate that Americans' peak automobile use (distance driven) occurs between the ages of 35 and 54. Thereafter, distances driven by men drop by around 12 per cent until age 64 and about 47 per cent from the peak after age 65; the corresponding figures for women are about 4 per cent and 64 per cent. Similar although probably not identical behaviour almost certainly occurs in other countries. Because demographers can forecast the size and age structures of populations accurately for at least a generation, an indicator of the likely tendency towards declining energy use from population ageing is easily calculated and combined or compared with other indicators which may not go in the same direction.

■ The problem of linking energy and non-energy policies also has a more general dimension. Governments need to be more alert to structural policies – especially subsidy and tax subsidy schemes – that may increase energy demands without properly allocating or at least considering the energy costs. Policies that subsidise home ownership or car ownership and use, spread people into more sparsely populated areas, or locate production far from suppliers or markets all are candidates for scrutiny. Not all should be challenged on energy grounds, but policy analysts should estimate their impacts. Governments wanting to design effective energy policies would do well to scour their inter-ministerial systems for knowledge of their own policies which may be at odds with energy objectives.

■ At the same time, governments need to track trends that may affect energy use and CO_2 emissions but seem remote from policy concerns, such as changes in the labour market or changes in family size and structure. Such analysis of social and demographic forces belongs in an indicators system and could in the end have considerable relevance.

■ *Monitoring progress* is the final step in the policy process. Indicators have a dual role here because monitoring involves not only tracking the effects of policies themselves but also watching for, spotting and evaluating spontaneous developments that could affect outcomes.

Policies About Indicators

■ Fully developed indicators systems do not now exist in most IEA countries and those which do need improvement. Yet all countries have in hand more data, more of the building blocks of such systems, than they now realise. Authorities and experts routinely form and use the indicators illustrated here; there is really little new about them. For most analysts, the task of developing the indicators means achieving more depth in the data, more frequent surveys of relevant activities and better links to non-energy data sources, which often exist within their own national systems. These steps have low marginal costs. As shown, some IEA countries understand the usefulness and relevance and are pursuing the approach. Non IEA countries also would find this approach useful as they are facing equally important challenges.

■ In a world obsessed with environmental and climate-change concerns, energy indicators have a critically important role to play. No other available method can offer the wealth of insight and interpretation that the indicators provide. Nor does any other approach bring the analyst into direct touch with the grass-roots decision-making process that, in the end, determines how much energy is used and in what ways. The indicators offer all the advantages of a bottom-up methodology – but are not limited to it. Properly handled, they can be made to yield aggregate data that are richer and more useful in guiding policy formation than the customary top-down approach achieves. With the help of indicators, authorities can attack the spread of CO_2 emissions with policies that respect patterns of human behaviour and so can gain political acceptance.

MAIN SALES OUTLETS OF OECD PUBLICATIONS
PRINCIPAUX POINTS DE VENTE DES PUBLICATIONS DE L'OCDE

AUSTRALIA – AUSTRALIE
D.A. Information Services
648 Whitehorse Road, P.O.B 163
Mitcham, Victoria 3132 Tel. (03) 9210.7777
Fax: (03) 9210.7788

AUSTRIA – AUTRICHE
Gerold & Co.
Graben 31
Wien I Tel. (0222) 533.50.14
Fax: (0222) 512.47.31.29

BELGIUM – BELGIQUE
Jean De Lannoy
Avenue du Roi, Koningslaan 202
B-1060 Bruxelles Tel. (02) 538.51.69/538.08.41
Fax: (02) 538.08.41

CANADA
Renouf Publishing Company Ltd.
5369 Canotek Road
Unit 1
Ottawa, Ont. K1J 9J3 Tel. (613) 745.2665
Fax: (613) 745.7660

Stores:
71 1/2 Sparks Street
Ottawa, Ont. K1P 5R1 Tel. (613) 238.8985
Fax: (613) 238.6041

12 Adelaide Street West
Toronto, QN M5H 1L6 Tel. (416) 363.3171
Fax: (416) 363.5963

Les Éditions La Liberté Inc.
3020 Chemin Sainte-Foy
Sainte-Foy, PQ G1X 3V6 Tel. (418) 658.3763
Fax: (418) 658.3763

Federal Publications Inc.
165 University Avenue, Suite 701
Toronto, ON M5H 3B8 Tel. (416) 860.1611
Fax: (416) 860.1608

Les Publications Fédérales
1185 Université
Montréal, QC H3B 3A7 Tel. (514) 954.1633
Fax: (514) 954.1635

CHINA – CHINE
Book Dept., China Natinal Publiations
Import and Export Corporation (CNPIEC)
16 Gongti E. Road, Chaoyang District
Beijing 100020 Tel. (10) 6506-6688 Ext. 8402
(10) 6506-3101

CHINESE TAIPEI – TAIPEI CHINOIS
Good Faith Worldwide Int'l. Co. Ltd.
9th Floor, No. 118, Sec. 2
Chung Hsiao E. Road
Taipei Tel. (02) 391.7396/391.7397
Fax: (02) 394.9176

CZECH REPUBLIC – RÉPUBLIQUE TCHÈQUE
National Information Centre
NIS – prodejna
Konviktská 5
Praha 1 – 113 57 Tel. (02) 24.23.09.07
Fax: (02) 24.22.94.33
E-mail: nkposp@dec.niz.cz
Internet: http://www.nis.cz

DENMARK – DANEMARK
Munksgaard Book and Subscription Service
35, Nørre Søgade, P.O. Box 2148
DK-1016 København K Tel. (33) 12.85.70
Fax: (33) 12.93.87

J. H. Schultz Information A/S,
Herstedvang 12,
DK – 2620 Albertslung Tel. 43 63 23 00
Fax: 43 63 19 69
Internet: s-info@inet.uni-c.dk

EGYPT – ÉGYPTE
The Middle East Observer
41 Sherif Street
Cairo Tel. (2) 392.6919
Fax: (2) 360.6804

FINLAND – FINLANDE
Akateeminen Kirjakauppa
Keskuskatu 1, P.O. Box 128
00100 Helsinki

Subscription Services/Agence d'abonnements :
P.O. Box 23
00100 Helsinki Tel. (358) 9.121.4403
Fax: (358) 9.121.4450

***FRANCE**
OECD/OCDE
Mail Orders/Commandes par correspondance :
2, rue André-Pascal
75775 Paris Cedex 16 Tel. 33 (0)1.45.24.82.00
Fax: 33 (0)1.49.10.42.76
Telex: 640048 OCDE
Internet: Compte.PUBSINQ@oecd.org

Orders via Minitel, France only/
Commandes par Minitel, France exclusivement :
36 15 OCDE

OECD Bookshop/Librairie de l'OCDE :
33, rue Octave-Feuillet
75016 Paris Tel. 33 (0)1.45.24.81.81
33 (0)1.45.24.81.67

Dawson
B.P. 40
91121 Palaiseau Cedex Tel. 01.89.10.47.00
Fax: 01.64.54.83.26

Documentation Française
29, quai Voltaire
75007 Paris Tel. 01.40.15.70.00

Economica
49, rue Héricart
75015 Paris Tel. 01.45.78.12.92
Fax: 01.45.75.05.67

Gibert Jeune (Droit-Économie)
6, place Saint-Michel
75006 Paris Tel. 01.43.25.91.19

Librairie du Commerce International
10, avenue d'Iéna
75016 Paris Tel. 01.40.73.34.60

Librairie Dunod
Université Paris-Dauphine
Place du Maréchal-de-Lattre-de-Tassigny
75016 Paris Tel. 01.44.05.40.13

Librairie Lavoisier
11, rue Lavoisier
75008 Paris Tel. 01.42.65.39.95

Librairie des Sciences Politiques
30, rue Saint-Guillaume
75007 Paris Tel. 01.45.48.36.02

P.U.F.
49, boulevard Saint-Michel
75005 Paris Tel. 01.43.25.83.40

Librairie de l'Université
12*a*, rue Nazareth
13100 Aix-en-Provence Tel. 04.42.26.18.08

Documentation Française
165, rue Garibaldi
69003 Lyon Tel. 04.78.63.32.23

Librairie Decitre
29, place Bellecour
69002 Lyon Tel. 04.72.40.54.54

Librairie Sauramps
Le Triangle
34967 Montpellier Cedex 2 Tel. 04.67.58.85.15
Fax: 04.67.58.27.36

A la Sorbonne Actual
23, rue de l'Hôtel-des-Postes
06000 Nice Tel. 04.93.13.77.75
Fax: 04.93.80.75.69

GERMANY – ALLEMAGNE
OECD Bonn Centre
August-Bebel-Allee 6
D-53175 Bonn Tel. (0228) 959.120
Fax: (0228) 959.12.17

GREECE – GRÈCE
Librairie Kauffmann
Stadiou 28
10564 Athens Tel. (01) 32.55.321
Fax: (01) 32.30.320

HONG-KONG
Swindon Book Co. Ltd.
Astoria Bldg. 3F
34 Ashley Road, Tsimshatsui
Kowloon, Hong Kong Tel. 2376.2062
Fax: 2376.0685

HUNGARY – HONGRIE
Euro Info Service
Margitsziget, Európa Ház
1138 Budapest Tel. (1) 111.60.61
Fax: (1) 302.50.35
E-mail: euroinfo@mail.matav.hu
Internet: http://www.euroinfo.hu//index.html

ICELAND – ISLANDE
Mál og Menning
Laugavegi 18, Pósthólf 392
121 Reykjavik Tel. (1) 552.4240
Fax: (1) 562.3523

INDIA – INDE
Oxford Book and Stationery Co.
Scindia House
New Delhi 110001 Tel. (11) 331.5896/5308
Fax: (11) 332.2639
E-mail: oxford.publ@axcess.net.in

17 Park Street
Calcutta 700016 Tel. 240832

INDONESIA – INDONÉSIE
Pdii-Lipi
P.O. Box 4298
Jakarta 12042 Tel. (21) 573.34.67
Fax: (21) 573.34.67

IRELAND – IRLANDE
Government Supplies Agency
Publications Section
4/5 Harcourt Road
Dublin 2 Tel. 661.31.11
Fax: 475.27.60

ISRAEL – ISRAËL
Praedicta
5 Shatner Street
P.O. Box 34030
Jerusalem 91430 Tel. (2) 652.84.90/1/2
Fax: (2) 652.84.93

R.O.Y. International
P.O. Box 13056
Tel Aviv 61130 Tel. (3) 546 1423
Fax: (3) 546 1442
E-mail: royil@netvision.net.il

Palestinian Authority/Middle East:
INDEX Information Services
P.O.B. 19502
Jerusalem Tel. (2) 627.16.34
Fax: (2) 627.12.19

ITALY – ITALIE
Libreria Commissionaria Sansoni
Via Duca di Calabria, 1/1
50125 Firenze Tel. (055) 64.54.15
Fax: (055) 64.12.57
E-mail: licosa@ftbcc.it

Via Bartolini 29
20155 Milano Tel. (02) 36.50.83

Editrice e Libreria Herder
Piazza Montecitorio 120
00186 Roma Tel. 679.46.28
Fax: 678.47.51

Libreria Hoepli
Via Hoepli 5
20121 Milano Tel. (02) 86.54.46
Fax: (02) 805.28.86

Libreria Scientifica
Dott. Lucio de Biasio 'Aeiou'
Via Coronelli, 6
20146 Milano Tel. (02) 48.95.45.52
Fax: (02) 48.95.45.48

JAPAN – JAPON
OECD Tokyo Centre
Landic Akasaka Building
2-3-4 Akasaka, Minato-ku
Tokyo 107 Tel. (81.3) 3586.2016
Fax: (81.3) 3584.7929

KOREA – CORÉE
Kyobo Book Centre Co. Ltd.
P.O. Box 1658, Kwang Hwa Moon
Seoul Tel. 730.78.91
Fax: 735.00.30

MALAYSIA – MALAISIE
University of Malaya Bookshop
University of Malaya
P.O. Box 1127, Jalan Pantai Baru
59700 Kuala Lumpur
Malaysia Tel. 756.5000/756.5425
Fax: 756.3246

MEXICO – MEXIQUE
OECD Mexico Centre
Edificio INFOTEC
Av. San Fernando no. 37
Col. Toriello Guerra
Tlalpan C.P. 14050
Mexico D.F. Tel. (525) 528.10.38
Fax: (525) 606.13.07
E-mail: ocde@rtn.net.mx

NETHERLANDS – PAYS-BAS
SDU Uitgeverij Plantijnstraat
Externe Fondsen
Postbus 20014
2500 EA's-Gravenhage Tel. (070) 37.89.880
Voor bestellingen: Fax: (070) 34.75.778

Subscription Agency/Agence d'abonnements :
SWETS & ZEITLINGER BV
Heereweg 347B
P.O. Box 830
2160 SZ Lisse Tel. 252.435.111
Fax: 252.415.888

NEW ZEALAND – NOUVELLE-ZÉLANDE
GPLegislation Services
P.O. Box 12418
Thorndon, Wellington Tel. (04) 496.5655
Fax: (04) 496.5698

NORWAY – NORVÈGE
NIC INFO A/S
Ostensjoveien 18
P.O. Box 6512 Etterstad
0606 Oslo Tel. (22) 97.45.00
Fax: (22) 97.45.45

PAKISTAN
Mirza Book Agency
65 Shahrah Quaid-E-Azam
Lahore 54000 Tel. (42) 735.36.01
Fax: (42) 576.37.14

PHILIPPINE – PHILIPPINES
International Booksource Center Inc.
Rm 179/920 Cityland 10 Condo Tower 2
HV dela Costa Ext cor Valero St.
Makati Metro Manila Tel. (632) 817 9676
Fax: (632) 817 1741

POLAND – POLOGNE
Ars Polona
00-950 Warszawa
Krakowskie Prezdmiescie 7 Tel. (22) 264760
Fax: (22) 265334

PORTUGAL
Livraria Portugal
Rua do Carmo 70-74
Apart. 2681
1200 Lisboa Tel. (01) 347.49.82/5
Fax: (01) 347.02.64

SINGAPORE – SINGAPOUR
Ashgate Publishing
Asia Pacific Pte. Ltd
Golden Wheel Building, 04-03
41, Kallang Pudding Road
Singapore 349316 Tel. 741.5166
Fax: 742.9356

SPAIN – ESPAGNE
Mundi-Prensa Libros S.A.
Castelló 37, Apartado 1223
Madrid 28001 Tel. (91) 431.33.99
Fax: (91) 575.39.98
E-mail: mundiprensa@tsai.es
Internet: http://www.mundiprensa.es

Mundi-Prensa Barcelona
Consell de Cent No. 391
08009 – Barcelona Tel. (93) 488.34.92
Fax: (93) 487.76.59

Libreria de la Generalitat
Palau Moja
Rambla dels Estudis, 118
08002 – Barcelona
(Suscripciones) Tel. (93) 318.80.12
(Publicaciones) Tel. (93) 302.67.23
Fax: (93) 412.18.54

SRI LANKA
Centre for Policy Research
c/o Colombo Agencies Ltd.
No. 300-304, Galle Road
Colombo 3 Tel. (1) 574240, 573551-2
Fax: (1) 575394, 510711

SWEDEN – SUÈDE
CE Fritzes AB
S–106 47 Stockholm Tel. (08) 690.90.90
Fax: (08) 20.50.21

For electronic publications only/
Publications électroniques seulement
STATISTICS SWEDEN
Informationsservice
S-115 81 Stockholm Tel. 8 783 5066
Fax: 8 783 4045

Subscription Agency/Agence d'abonnements :
Wennergren-Williams Info AB
P.O. Box 1305
171 25 Solna Tel. (08) 705.97.50
Fax: (08) 27.00.71

Liber distribution
Internatinal organizations
Fagerstagatan 21
S-163 52 Spanga

SWITZERLAND – SUISSE
Maditec S.A. (Books and Periodicals/Livres et périodiques)
Chemin des Palettes 4
Case postale 266
1020 Renens VD 1 Tel. (021) 635.08.65
Fax: (021) 635.07.80

Librairie Payot S.A.
4, place Pépinet
CP 3212
1002 Lausanne Tel. (021) 320.25.11
Fax: (021) 320.25.14

Librairie Unilivres
6, rue de Candolle
1205 Genève Tel. (022) 320.26.23
Fax: (022) 329.73.18

Subscription Agency/Agence d'abonnements :
Dynapresse Marketing S.A.
38, avenue Vibert
1227 Carouge Tel. (022) 308.08.70
Fax: (022) 308.07.99

See also – Voir aussi :
OECD Bonn Centre
August-Bebel-Allee 6
D-53175 Bonn (Germany) Tel. (0228) 959.120
Fax: (0228) 959.12.17

THAILAND – THAÏLANDE
Suksit Siam Co. Ltd.
113, 115 Fuang Nakhon Rd.
Opp. Wat Rajbopith
Bangkok 10200 Tel. (662) 225.9531/2
Fax: (662) 222.5188

TRINIDAD & TOBAGO, CARIBBEAN TRINITÉ-ET-TOBAGO, CARAÏBES
Systematics Studies Limited
9 Watts Street
Curepe
Trinadad & Tobago, W.I. Tel. (1809) 645.3475
Fax: (1809) 662.5654
E-mail: tobe@trinidad.net

TUNISIA – TUNISIE
Grande Librairie Spécialisée
Fendri Ali
Avenue Haffouz Imm El-Intilaka
Bloc B 1 Sfax 3000 Tel. (216-4) 296 855
Fax: (216-4) 298.270

TURKEY – TURQUIE
Kültür Yayinlari Is-Türk Ltd.
Atatürk Bulvari No. 191/Kat 13
06684 Kavaklidere/Ankara
Tel. (312) 428.11.40 Ext. 2458
Fax : (312) 417.[illegible]

Dolmabahce Cad. No. 29
Besiktas/Istanbul Tel. (212) 260 7188

UNITED KINGDOM – ROYAUME-UNI
The Stationery Office Ltd.
Postal orders only:
P.O. Box 276, London SW8 5DT
Gen. enquiries Tel. (171) 873 0011
Fax: (171) 873 8463

The Stationery Office Ltd.
Postal orders only:
49 High Holborn, London WC1V 6HB

Branches at: Belfast, Birmingham, Bristol, Edinburgh, Manchester

UNITED STATES – ÉTATS-UNIS
OECD Washington Center
2001 L Street N.W., Suite 650
Washington, D.C. 20036-4922 Tel. (202) 785.6323
Fax: (202) 785.0350
Internet: washcont@oecd.org

Subscriptions to OECD periodicals may also be placed through main subscription agencies.

Les abonnements aux publications périodiques de l'OCDE peuvent être souscrits auprès des principales agences d'abonnement.

Orders and inquiries from countries where Distributors have not yet been appointed should be sent to: OECD Publications, 2, rue André-Pascal, 75775 Paris Cedex 16, France.

Les commandes provenant de pays où l'OCDE n'a pas encore désigné de distributeur peuvent être adressées aux Éditions de l'OCDE, 2, rue André-Pascal, 75775 Paris Cedex 16, France.

12-1996

OECD PUBLICATIONS, 2 rue André-Pascal, 75775 PARIS CEDEX 16
PRINTED IN FRANCE - Photos PIX
(61 97 33 1 P) ISBN 92-64-15690-9 – 1997